Neglected and Underutilized Crops

This book provides a comprehensive review of the potential of neglected and underutilized crops in the prevention and treatment of various cancers. It explores various aspects of these crops and their bioactive compounds, shedding light on their role in health and nutrition. The initial chapters discuss neglected and underutilized crops, emphasizing their relevance in human disease and cancer management. These chapters also examine the bioavailability and formulation strategies of bioactive compounds derived from these crops.

This book further delves into the isolation, availability, and therapeutic uses of these bioactive compounds, emphasizing their significant role in modern medicine. It also reviews the potential of the Indian gooseberry, buckwheat, quinoa, and sea buckthorn in cancer treatment and management. Additionally, this book discusses neglected medicinal plants of the northwestern Himalayas, showcasing their potential applications in both traditional and modern medicine. It also highlights the importance of sustainable cultivation and conservation efforts to ensure the long-term availability of these crops. This book is of particular interest to scientists, researchers, and academics in fields such as agriculture, biochemistry, nutrition, and medicine.

Neglected and Underutilized Crops

Unveiling Potential for Cancer Drug Discovery

Edited by
Muzafar Ahmad Macha, Sajad Majeed Zargar,
Ajaz A. Bhat, and Shahab Uddin

CRC Press
Taylor & Francis Group
Boca Raton London New York

CRC Press is an imprint of the
Taylor & Francis Group, an **informa** business

First edition published 2025
by CRC Press
2385 NW Executive Center Drive, Suite 320, Boca Raton FL 33431

and by CRC Press
4 Park Square, Milton Park, Abingdon, Oxon, OX14 4RN

CRC Press is an imprint of Taylor & Francis Group, LLC

ISBN: 9781032762692 (hbk)
ISBN: 9781032762708 (pbk)
ISBN: 9781003477792 (ebk)

DOI: 10.1201/9781003477792

Typeset in Times
by codeMantra

Contents

About the Editors

Dr. Muzafar Ahmad Macha is currently serving as an assistant professor and a Ramalingaswami Fellow at the Watson-Crick Centre for Molecular Medicine, Islamic University of Science and Technology (IUST), India. He earned his master's degree in biochemistry from Jamia Hamdard in New Delhi, India. In 2005, he embarked on his doctoral program, focusing on head and neck cancers (HNCs), at the Department of Biochemistry at the All India Institute of Medical Sciences (AIIMS) in New Delhi, India. In 2011, he joined the Department of Biochemistry and Molecular Biology (BMB) at the University of Nebraska Medical Center (UNMC), USA, as a postdoctoral fellow. During his tenure, he established an independent HNC Research Group and conducted research on the identification of deregulated signaling pathways and novel therapeutic targets. Following his promotion to assistant professor in the BMB at UNMC, he developed numerous in vitro cell lines, genetically engineered mouse models (GEMMs) of head and neck squamous cell carcinoma (HNSCC), and unique therapeutic modalities. Dr. Macha has published his work in many peer-reviewed journals. He has also received numerous awards, including the Young Investigator Award; the Scholarship Award; the Scholar-in-Training Award; the Best Performance Stipend Award for the years 2012 and 2013; the Gita Mittal Award/gold medal for the best basic cancer research in Ph.D. from AIIMS, New Delhi; and a gold medal for his master's degree from Jamia Hamdard, New Delhi. More recently, he has been honored with two prestigious fellowships: the Ramalingaswami Fellowship and the Ramanujan Fellowship from the Government of India.

Dr. Sajad Majeed Zargar is currently an associate professor at Sher-e-Kashmir University of Agricultural Sciences and Technology of Kashmir (SKUAST-Kashmir), India, and a visiting professor at the University of Padova, Italy. He was previously a visiting professor at the Nara Institute of Science and Technology, Japan. He has worked as an assistant professor at SKUAST-Jammu and Baba Ghulam Shah Badshah University, Rajouri (BGSB), India. He has also worked as a scientist at Advanta India Limited, Hyderabad, India, and TERI (The Energy and Resources Institute), New Delhi, India. Dr. Zargar is the recipient of a CREST overseas fellowship from DBT, India; a World Bank funded (NAHEP) overseas fellowship, a Goho grant from the Government of Japan; and an Erasmus fellowship from the European Union. He has received several awards for his work and research. He is also a member and representative of INPPO (International Plant Proteomics Organization). He chaired a session at the 3rd INPPO World Congress 2018 held at the University of Padova, Italy. His editorial activities and scientific memberships include publishing research and review articles in international journals and serving as a reviewer for reputed journals. He is the in-charge scientist of Genomics and Proteomics Labs at the Division of Plant Biotechnology, SKUAST-Kashmir, India.

Dr. Ajaz A. Bhat works as a senior staff scientist at Sidra Medicine, Qatar. He completed his Ph.D. at the All India Institute of Medical Sciences (AIIMS), New Delhi, focusing on vaccine immunology. After obtaining his Ph.D., he moved to Vanderbilt University Medical School (VUMS), Nashville, TN, USA, for a postdoctoral fellowship. During his 11 years of post-Ph.D. training, he gained extensive experience in the pathobiological importance of tight junction proteins (claudins) in colorectal cancer and the role of bile acid-induced APE1/Ref1 in the pathogenesis of esophageal cancer. His current research focuses on drug discovery, signaling networks in cancer, cancer metabolism, cancer genomics, obesity, and diabetes. He has published his research in journals of high repute. He has also attended numerous international meetings, conferences, and symposia to present his research results and has received awards from the Immunology Society and the American Association of Cancer Research. Additionally, he was recognized as an outstanding researcher by Sidra Medicine. He is an active member of several immunology and cancer societies and serves on the editorial boards and review committees of numerous peer-reviewed journals.

Prof. Shahab Uddin currently holds the position of Principal Rsearch Scientist, Professor of Cancer Biology and leads the Molecular Pathophysiology Core Lab at the Translational Research Institute, Hamad Medical Corporation, Qatar. With over 20 years of experience in translational research, he has published more than 200 peer-reviewed research papers. Dr. Uddin has been involved as a principal investigator (PI) and co-investigator (CO-I) in numerous translational research projects funded by organizations, such as King Abdulaziz Science and Technology in Saudi Arabia, King Faisal Hospital and Research Center, and the Qatar National Research Fund. He has also managed grants from the American Cancer Society and the Wellcome Research Foundation. Additionally, he serves as a reviewer and editorial board member for several international journals.

List of Contributors

Hussaina Banu Buhari Malkhan Ali
Department of Applied Sciences, College of Arts & Sciences
Abu Dhabi University
Abu Dhabi, United Arab Emirates

Mudasir Bashir
Department of Physiology
Graphic Era Institute of Medical Sciences (GEIMS)
Dehradun, India

Zaffar Bashir
Department of Microbiology
University of Kashmir
Srinagar, India

M. Ashraf Bhat
Division of Plant Biotechnology
Sher-e-Kashmir University of Agricultural Sciences and Technology of Kashmir (SKUAST-K)
Srinagar, India

Waseem Ali Dar
Krishi Vigyan Kendra/Extension Training Centre, Pulwama
Sher-e-Kashmir University of Agricultural Sciences and Technology of Kashmir
Srinagar, India

Rajkumari Jashmi Devi
Plant Bioresources Division
Institute of Bioresources and Sustainable Development (IBSD)
Imphal West, India

Yengkhom Linthoingambi Devi
Plant Bioresources Division
Institute of Bioresources and Sustainable Development (IBSD)
Imphal West, India

Bashir Ah Ganai
Centre of Research for Development
University of Kashmir
Srinagar, India

Aadil Gani
Division of Plant Biotechnology
Sher-e-Kashmir University of Agricultural Sciences and Technology of Kashmir (SKUAST-K)
Srinagar, India

Shahnawaz Hassan
Department of Environmental Science
University of Kashmir
Srinagar, India

Pervaiz Iqbal
Department of Mathematics & Actuarial Science
B.S. Abdur Rahman Crescent Institute of Science & Technology
Chennai, India

Moien Rasheed Lone
David Geffen School of Medicine, Center for Health Sciences
University of California
Los Angeles, California

Reetika Mahajan
Department of Medicinal and Biological Chemistry, College of Pharmacy and Pharmaceutical Sciences
University of Toledo
Toledo, Ohio

Firdose A. Malik
College of Temperate Sericulture, Mirgund
Sher-e-Kashmir University of Agricultural Sciences and Technology of Kashmir (SKUAST-K)
Srinagar, India

Mudasir A. Mir
Division of Plant Biotechnology
Sher-e-Kashmir University of Agricultural Sciences and Technology of Kashmir (SKUAST-K)
Srinagar, India

J. A. Mugloo
Krishi Vigyan Kendra/Extension Training Centre, Pulwama
Sher-e-Kashmir University of Agricultural Sciences and Technology of Kashmir
Srinagar, India

Shalini Mukherjee
Scientific Writer
Aabir Bio-Services Foundation
Phulwarisharif, Patna, India

Md. Kausar Neyaz
Aabir Bio-services Foundation
Patna, India

Mohamad Maqbool Pakhtoon
Proteomics Laboratory, Division of Plant Biotechnology
Sher-e-Kashmir University of Agricultural Sciences and Technology of Kashmir (SKUAST-K)
Srinagar, India
and
Department of Life Sciences
Rabindranath Tagore University
Bhopal, India
Aabir Bio-services Foundation
Patna, India

Asif R. Rather
College of Temperate Sericulture, Mirgund
Sher-e-Kashmir University of Agricultural Sciences and Technology of Kashmir (SKUAST-K)
Srinagar, India

Sabreena
Department of Environmental Science
University of Kashmir
Srinagar, India

Geetanjali Sageena
Division of Basic Medical Sciences
Indian Council of Medical Research
New Delhi, India

Sheikh Imran Sayeed
Department of Physiology
Government Medical College Srinagar
Srinagar, India

Lubna Shafi
Department of Mathematics & Actuarial Science
B.S. Abdur Rahman Crescent Institute of Science & Technology
Chennai, India

Zahoor A. Shah
Department of Medicinal and Biological Chemistry, College of Pharmacy and Pharmaceutical Sciences
University of Toledo
Toledo, Ohio

Dr. Khushboo Sharma
Department of Zoology
Banaras Hindu University
Varanasi, UP, India

Manish Sharma
Department of Zoology
University of Allahabad
Allahabad, India

Faheem Shehjar
Department of Medicinal and Biological Chemistry, College of Pharmacy and Pharmaceutical Sciences
University of Toledo
Toledo, Ohio

Biseshwori Thongam
Plant Bioresources Division
Institute of Bioresources and Sustainable Development (IBSD)
Imphal West, India

1 Introduction to Neglected and Underutilized Crops

Mudasir A. Mir, M. Ashraf Bhat, Aadil Gani, Firdose A. Malik, and Asif R. Rather
Sher-e-Kashmir University of Agricultural Sciences and Technology of Kashmir (SKUAST-K), Srinagar, India

Zaffar Bashir
University of Kashmir, Srinagar, India

1.1 INTRODUCTION

The world population is growing at an alarming rate, and the current global human population is approximately 7.9 billion which is anticipated to touch 9.7 billion by 2050. Moreover, food crops like wheat, rice, and maize that constitute a major part of the food supply seem scanty to meet the food demand of the growing populations [1,2]. Aside from uncontrolled population expansion, different biotic and abiotic stresses limit plant growth and development, which has a detrimental impact on agricultural yield and production. These include variable temperatures, severe and extended droughts, soil degradation, excessive salinity, floods, and an increase in disease and pests. As a result, it is critical to investigate the existing genetic heterogeneity of plants in order to increase the production and distribution of food while reducing over-dependence on key food crops and promoting global food security [3,4].

At least 720 million people (9.9% of the global population) were hungry in 2020, an evident increase from prior years and the highest percentage of overall population since 2010 [5]. The constantly changing climate, increase in global population, and the COVID-19 pandemic would have a proportionate effect on people toward hunger. Accomplishing Sustainable Development Goal (SDG) 2 ("Zero Hunger") may be the puzzle even by 2030 [6], but the most versatile and formidable tool for achieving Zero Hunger could be agricultural diversification [7]. However, the present way of agriculture and food systems [8] has put limitations on production diversity that resulted in malnutrition due to imbalanced diets.

In Asia and the Pacific region, people usually lack wholesome nutrition in their diet because they grow and consume only few major staple crops like rice. Hence, they are far away from the recommended nutrition intake [9]. According to the Food and Agriculture Organization of the United Nations (FAO), relying on only a few crops has a negative impact on ecological balance, food varieties, and human well-being. Monoculture and

DOI: 10.1201/9781003477792-1

monotony in food would escalate the hazard of micronutrient insufficiencies [10]. The consumption of calorie-rich and homogenized diets having low nutritional profile may also result in a rise in concealed hunger, which is a serious concern to human health [11,12,58,60]. Dietary diversification is an inexpensive, cost-effective, and long-term strategy for reducing malnutrition and hunger, while diversification of production promotes the provision of a wide range of highly nutritious foods and addresses the effects of climate change [13,62]. Therefore, there can be no better way than to utilize neglected and underutilized (NUS) species/crops to achieve high nutrition profile and food security and to combat the rising hunger, malnutrition, diseases, etc.

1.2 WHAT ARE NEGLECTED AND UNDERUTILIZED (NUS) CROPS?

Crops are classified into two groups, i.e., staple and underused. The majority of underutilized crops are wild or semi-domesticated species suited to local settings, whereas staple crops are domesticated, highly commercialized, and are the main food considered for consumption in a particular place. The term "underutilized" may be defined as the neglecting of a species by native and global research groups. It is not necessary that a crop underutilized in one place may not be commercial at another place. Modernization of agriculture, extensive monoculture, and the promotion of high-yielding cultivars are all factors that may be a viable reason that underutilized crops have not gained much attention at the global level, while for rice cultivation, more emphasis has been laid down by governments, as seen by the rice self-sufficiency policy which indicates discrimination toward underutilized crops. The idea of NUS crops as "poor food" has stigmatized it culturally, having a negative effect on its production and consumption [14]. However, in terms of food security and climate change, these crops play a promising and vital role, while in developing nations, improving the living conditions of low-income families and customers is a priority [14,15]. Crop species that are underutilized and have the necessary and beneficial nutrient profile can fill a variety of positions in food production systems and may be well suited to cost-effective systems in areas where resources for building a successful high-through put agricultural system are limited [16,17]. Globally, NUS crops may vary from place to place. However, Li and Siddique have categorized some future smart and underutilized crops for eight countries in Asia which are given in Table 1.1 [13]. This categorization will help the researchers to prioritize these crops for better agricultural sustainability (Table 1.2).

1.3 IMPORTANCE OF NUS

Neglected and underused species offer us greater ways and chances to tackle issues like poverty, starvation, and malnourishment, and their adoption into farming systems might result in nutrient-rich, climate-resilient, and resource-efficient agriculture. Thus, they have a significant economic importance [13].

1.3.1 FIGHT FOOD INSECURITY

The present global food scenario is fragile in terms of food security since it relies on a small number of basic foods such as rice, wheat, and maize [18]. Current farming

TABLE 1.1
Future Smart and Underutilized Crops

Cereals	Roots and Tubers	Pulses	Fruits and Vegetables	Nuts, Seeds, and Spices
Buckwheat	Taro	Grass pea	Drumstick	Linseed
Tartary buckwheat	Swamp taro	Faba bean	Chayote	Walnut
Foxtail millet	Purple yam	Cowpea	Fenugreek	Nepali butter tree
Finger millet	Fancy yam	Mung bean	Snake gourd	Perilla
Sorghum	Elephant's foot yam	Black gram	Pumpkin	Nepali pepper
Amaranth	Sweet potato	Rice bean	Roselle	-
Grain amaranth	-	Lentil	Indian gooseberry	-
Quinoa	-	Horse gram	Jackfruit	-
Specialty rice	-	Soybean	Wood apple	-

Source: Li and Siddique (2018).

TABLE 1.2
Medicinal and Economic Use of Underutilized Crops

S.No.	Plant Name	Importance (Medicinal and Economic)
1	Amaranth (*Amaranthus* spp.)	Astringent, for skin problems, hypolipidemic, antihyperglycemic, against microorganisms, neuroprotective or antidepressant, anti-inflammatory, antioxidant, and anti-cancer [7]
2	African nightshade (*Solanum scabrum*)	Cooked vegetable, treat diarrhea and ulcers, eye infections, and jaundice [24]
3	Asian (*Solanum melongena*)	Against asthma, allergic rhinitis, nasal catarrh, skin infections, rheumatic disease, swollen joint pains, gastro-esophageal reflux disease, constipation, and dyspepsia [16]
4	Jute mallow (*Corchorus olitorius*)	Against malaria or typhoid fever, hypoglycemic [1]
5	Drumstick tree (*Moringa oleifera*)	Antitumor, antipyretic, antiepileptic, anti-inflammatory, antiulcer, antispasmodic, diuretic, antihypertensive, cholesterol lowering, antioxidant, antidiabetic, hepatoprotective, antibacterial, and antifungal activities [5]
6	Bitter gourd (*Momordicacharantia*)	Anti-helmintic, treatment of leprosy, piles and jaundice; treatment of ringworm, bowel movement, cough, congestion and chest pain, treat pneumonia and leukorrhea, used for jaundice, piles, leprosy, rheumatism, gout, diabetes, hydrophobia; treatment of malarial fevers anti-helminthic [56]
7	African eggplant (*Solanum aethiopicum*)	Antioxidant, analgesic, antidiabetic, cardio-protective, antipyretic, and hypolipidemic [53]

(*Continued*)

TABLE 1.2 (*Continued*)
Medicinal and Economic Use of Underutilized Crops

S.No.	Plant Name	Importance (Medicinal and Economic)
8	Water spinach (*Ipomoea aquatica*)	Treatment of liver diseases, antidiabetes, mental illness, anthelmintic, anti-depression, antiepileptic, antimicrobial, and anti-inflammatory [29]
9	*Brassica oleracea var. alboglabra*	Anti-cancer, antidiabetes, anti-asthma, and antioxidant [49]
10	Winged bean (*Psophocarpus tetragon olobus*)	Antioxidants [27]
11	Roselle (*Hibiscus sabdariffa*)	Antiseptic, diuretic, purgative, sedative, and emollient [42]
12	Chinese kale (*Brassica chinensis L.*)	Antioxidant [54]
13	Malabar spinach (*Basella alba*)	Anti- hypertension, antifungal, anticonvulsant, analgesic, anti-inflammatory, and androgenic activities and for the treatment of anemia [52]

practices emphasize monocultures with heavy inputs, which make operations easier but obviously jeopardize food security [19]. The majority of cultivated crops have a low genetic foundation throughout their germplasm pools, making them vulnerable to pests, diseases, and abiotic stressors [20]. For example, the Great Irish Potato Famine, which commenced in 1845 and continued for six years, killed roughly two-fifths of the population of Ireland (more than a million people), apart from forcing a million to evacuate the nation, has left an ominous lesson in history. Potato blight, *Phytophthora infestans*, was the cause of the famine, the pathological state that decimated whole crop potatoes across Europe. In Ireland, the illness had a disproportionate impact since three-quarters of the population relied completely on potato for nourishment [21]. Since the 17th century, impoverished Irish smallholders have grown potatoes as a primary diet, as potato production is substantially greater than grain output. In Ireland, however, potatoes were largely of one kind, the Irish Lumper. After the fungal infectious disease onset, it had devastating effects on potato plants, although elsewhere in Europe the effect on potato fields was much less since there were more types of potatoes in the field, as well as a greater variety of other crops. Exploring and rediscovering neglected crops in a similar way might lessen the risk of over-dependence on the limited big staple crops. Agronomic sustainability is described as a well-functioning, mutually beneficial link between agriculture and nature that includes three genetic levels of variation: intraspecific, interspecific, and agroecosystems [22,23].

Marginalizing underutilized crops endangers agrobiodiversity and may pose a threat to the food system sustainability. However, neglected and underutilized crop species that are underused may have a favorable impact on sustainable agriculture by significantly decreasing the requirement for external chemical inputs, such as pesticides and inorganic fertilizers [24]. Introducing underutilized crops in a farming system may not only reduce pest and disease buildup but also increase income through an organic way when grown in synchrony or in rotation with main crops.

1.3.2 Reduce Malnutrition

"Hidden hunger" in simple terms means the consumption of foods that have enough amount of calories but low or inadequate minerals and vitamin content. It is a threatening and lethal factor that has been affecting the human health of both developing and developed countries since ages [25]. The reliance on only a few main basic crops rich in carbohydrates contributes to hidden hunger [26]. Since the start of the Green Revolution, the main focus of agricultural research was on raising crop yields to meet food demands for the world's rapidly growing population. However, compromised nutritional value and adverse effect of the diet deficient in essential nutrients such as minerals and vitamins are clearly witnessed among people relying on selected crops. Crop species that have been neglected or underused have a lot of possibilities to converse the trend of hidden hunger. They frequently have greater quantities of important micronutrients, phytochemicals (such as flavonoids, minerals, and vitamins), and favorable macronutrient status than their more prevalent commercial staple crop counterparts (energy, fat, protein, and carbohydrates) [27,59].

1.3.3 Nutritional Value

Underutilized crops are considered as nutrient-dense food as they contain a high amount of calories, vitamins, minerals, and valuable bioactive compounds that have a direct and visible effect on human health. There are a number of NUS crops that may be substituted during the failure of main crop due to some environment pressures and abiotic and biotic stresses. Some of the underutilized crops with their nutritional status are quinoa, millets, Bambara groundnut, lentil, etc. Quinoa is a nutritionally rich underutilized crop that was brought to the foreground after the declaration by the General Assembly of United Nations as the International Year of Quinoa in 2014. It possesses double protein, four times more iron, five times more dietary fiber, and 23 times more folate (folic acid) than that in rice [28]. Lentils have lot of micronutrients, especially iron (Fe), selenium (Se), and zinc (Zn) [29, 61]. Millets are sometimes referred as "high-energy" cereals because they contain more vitamin A, protein, and oil than that in maize [30]. Millets are a viable and ideal crop for addressing nutritional issues encountered by people, particularly by mountain populations, due to vitamin A deficiency in main food crops [31]. Pearl millet has enough calcium, iron, riboflavin, folic acid, and zinc than maize or rice, as well as more micronutrients (excluding calcium) [32]. Bambara groundnut seeds contain 61%–69% carbohydrate, 17%–27% protein, 3.6%–7.4% fat, and 3.3%–6.4% fiber, making them a rich nutritional source [33,34]. It also has 5.1–9.0 mg potassium (K), 2.9–10.6 mg sodium (Na) per 100 g, 95.5–99.0 mg calcium (Ca), and a significant quantity of zinc (20.98 1.07 mg/100 g), which has been linked to a lower risk of prostate cancer in males [35]. In comparison to soybeans, groundnut, and cowpea protein contents (64%, 74%, and 65%, respectively), the seeds of Bambara groundnut showed maximum percentage (80%) of better class of amino acids like leucine, arginine, lysine, methionine, and valine and may potentially complement foods that are lacking in essential amino acids [35].

1.3.4 Health Benefits

The health benefits of neglected and underused species are enormous (Figure 1.1). Wijesuriya *et al.* conducted an empirical investigation in Sri Lanka to see how a Fe-rich lentil diet affected Fe-deficient anemic youngsters [36]. These data suggest that foods that included lentils in children's meals can assist to lower the prevalence of anemia in Sri Lankan children by increasing their Fe nutritional status, even in the short term, and therefore has a possible impact in other populations. Similarly, *Pachyrhizus erosus*, commonly known as jicama Mexican turnip contributes numerous health benefits pictorially shown below [37]. Besides the above species, there are many other plants like *Artemisia sp.* and *Taraxacum sp.* with diverse health effects.

1.4 CHEMICAL CONSTITUENTS OF NUS

Underutilized crop species, especially medicinal plants, possess a huge number of bioactive chemical constituents identified as alkaloids, glycosides, flavonoids, phenolics, saponins, tannins, essential oils, and steroids. Alkaloids are nitrogenous chemicals with anti-cancer and immune-stimulant properties. They include highly active

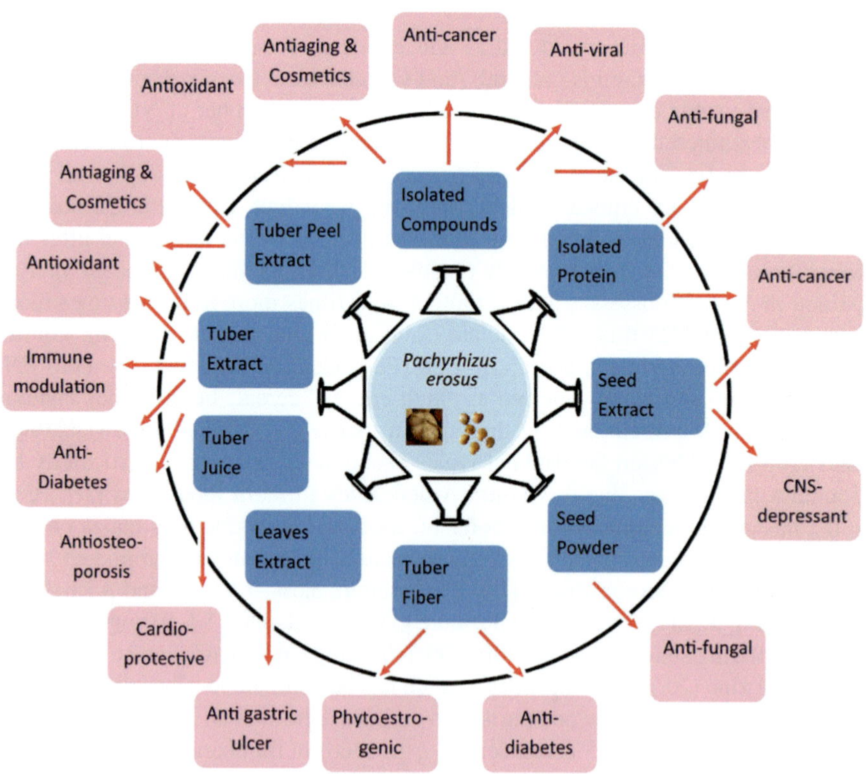

FIGURE 1.1 Different extracts and forms of *Pachyrhizus erosus* that have been found to possess diverse biological activity as per literature.

nitrogenous molecules [38]. Caffeine, theophylline, and theobromine are purine alkaloids found in plants that serve as stimulants improving cardiac activity, vasodilation, and metabolism [39].

a. Alkaloids: Some alkaloids of Madagascar periwinkle (*Catharanthus roseus* (L.) G. Don) are utilized to make medications with anti-cancer characteristics, such as vincristine and vinblastine, which are used to treat leukemia [40,41].
b. Cyanogenic glycosides: Some plants, such as elderberry and wild cherry, contain hydrogen cyanide, which works as a combating mechanism and relaxes muscle but can be hazardous to human health if taken in large quantities. Amygdalin found in bitter almonds is an example of cyanogenic glycoside [42,62]. The chemicals in elderberry and wild cherry are used to alleviate coughs [43].
c. Flavonoids: Flavonoids are secondary polyphenol metabolites that are abundant in nature and have a wide spectrum of therapeutic activities [44]. Plants have yielded more than 9,000 flavonoids till now, including antiviral, anti-inflammatory, cardio-protective, anti-cancer, anti-aging, and other characteristics [44]. Flavonoids, such as quercetin and kaempferol, may be found in a variety of plants, including vegetables [45].

1.5 ECONOMIC IMPORTANCE

Amaranth (*Amaranthus* spp.), African nightshade (*Solanum scabrum*), Asian (*Solanum melongena*), jute mallow (*Corchorus olitorius*), drumstick tree (*Moringaoleifera*), bitter gourd (*Momordica charantia*), African (*Solanum aethiopicum*) eggplant, water spinach (*Ipomoea aquatica*), (*Brassica oleracea var. alboglabra*), winged bean (*Psophocarpus tetragonolobus*), roselle (*Hibiscus sabdariffa*), Chinese kale, Malabar spinach (*Basella alba*), slippery cabbage (*Abelmoschus manihot*), and edible rape (*Brassica napus*), Vigna, have a high market significance and can thus have substantial impact on agriculture-based families [37,46,51,55, 57–62]. Value addition guarantees that superior quality products are released into the market and meet customer demands and expectations by providing proper production systems and postharvest practices. Consumer research on the purchasing and consumption of kale (*B. oleracea*) in Nairobi area of Kenya indicated that nutritional, aesthetic, and security properties are the most important to urban kale consumers [39,47]. According to an estimate, the greatest pay for green vegetables was found in lavish departmental stores (68%), followed by green markets (39%), retail stores (34%), and street markets. Selected indigenous vegetables have become a progressively fascinating and appealing diet for rich folklore in Eastern Africa and Southeast Asia and are steadily coming out of the underused group into the marketable mainstream [48]. As a result, the formal seed business is being revolutionized [49], as the yield of many of these crop varieties is simply too low for substantial commercial rewards [50]. Underutilized crops are thus very significant to leverage the economy of local populations as they harbor superior chemical fingerprints that needed to be exploited for the sustainability of agricultural produce.

1.6 CONCLUSION

Exploration of neglected and underutilized (NUS) crops to identify those crops having necessary and beneficial nutrient profile to tackle issues like poverty, starvation, and malnourishment elegantly must be the future goal of scientists. Also, neglected and underutilized crops have the potential to boost the local economy, as they harbor rich chemical composition that is needed to be exploited for the sustainability of the agricultural sector.

REFERENCES

1. Adebo, H. O., Ahoton, L. E., Quenum, F. J., Adoukonou-Sagbadja, H., Bello, D. O., & Chrysostome, C. A. (2018). Ethnobotanical knowledge of jute (Corchorus olitorius L.) in Benin. *Journal of Medicinal Plants*, 26(1), 1–11.
2. Adhikari, L., Hussain, A., & Rasul, G. (2017). Tapping the potential of neglected and underutilized food crops for sustainable nutrition security in the mountains of Pakistan and Nepal. *Sustainability*, 9(2), 291.
3. Akinola, R., Pereira, L. M., Mabhaudhi, T., De Bruin, F. M., & Rusch, L. (2020). A review of indigenous food crops in Africa and the implications for more sustainable and healthy food systems. *Sustainability*, 12(8), 3493.
4. Anaya, A. L., Cruz-Ortega, R., & Waller, G. R. (2006). Metabolism and ecology of purine alkaloids. *Frontiers in Bioscience-Landmark*, 11(3), 2354–2370.
5. Anwar, F., Latif, S., Ashraf, M., & Gilani, A. H. (2007). Moringa oleifera: a food plant with multiple medicinal uses. *Phytotherapy Research: An International Journal Devoted to Pharmacological and Toxicological Evaluation of Natural Product Derivatives*, 21(1), 17–25.
6. AOCC. The African Orphan Crops Consortium. 2019. Available online: https://africanorphancrops.org/ (accessed on 10 January 2019).
7. Baraniak, J., & Kania-Dobrowolska, M. (2022). The dual nature of amaranth-functional food and potential medicine. *Foods*, *11*(4), 618.
8. Barrales-Cureño, H. J., Reyes, C. R., García, I. V., Valdez, L. G. L., De Jesús, A. G., Cortés Ruíz, J. A., … Espinoza Perez, J. (2019). Alkaloids of pharmacological importance in Catharanthus roseus. *Alkaloids—Their Importance in Nature and Human Life*, 1, 18.
9. Bongaarts, J. (2021). FAO, IFAD, UNICEF, WFP and WHO The State of Food Security and Nutrition in the World 2020. Transforming food systems for affordable healthy diets FAO, 2020, 320 p.
10. Bruulsema, T. W., Heffer, P., Welch, R. M., Cakmak, I., & Moran, K. (2012). Fertilizing crops to improve human health: a scientific review. *Better Crops*, 2, 96.
11. Burchi, F., Fanzo, J., & Frison, E. (2011). The role of food and nutrition system approaches in tackling hidden hunger. *International Journal of Environmental Research and Public Health*, 8(2), 358–373.
12. Chankaew, S., Isemura, T., Naito, K., Ogiso-Tanaka, E., Tomooka, N., Somta, P., … Srinives, P. (2014). QTL mapping for salt tolerance and domestication-related traits in Vigna marina subsp. oblonga, a halophytic species. *Theoretical and Applied Genetics*, 127(3), 691–702.
13. Chanyalew, S., Ferede, S., Damte, T., Fikre, T., Genet, Y., Kebede, W., … Assefa, K. (2019). Significance and prospects of an orphan crop tef. *Planta*, 250(3), 753–767.
14. Dawson, I. K., Powell, W., Hendre, P., Bančič, J., Hickey, J. M., Kindt, R., … Jamnadass, R. (2019). The role of genetics in mainstreaming the production of new and orphan crops to diversify food systems and support human nutrition. *New Phytologist*, 224(1), 37–54.

15. DuVal, A., Mijatovic, D., & Hodgkin, T. (2019). The Contribution of biodiversity for food and agriculture to the resilience of production systems–*Thematic Study for The State of the World's Biodiversity for Food and Agriculture*. FAO, Rome. 85 pp. Licence: CC BY-NC-SA 3.0 IGO.
16. Edeke, A., Uchendu, N., Omeje, K., & Odiba, A. S. (2021). Nutritional and pharmacological potentials of Solanum melongena and Solanum aethiopicum fruits. *Journal of Phytopharmacology*, 10, 61–67.
17. Fairulnizal, M. N., Norhayati, M. K., Zaiton, A., Norliza, A. H., Rusidah, S., Aswir, A. R., … Zainuldin, T. M. (2015). Nutrient content in selected commercial rice in Malaysia: an update of Malaysian food composition database. *International Food Research Journal*, 22(2), 768.
18. Fanzo, J., Hunter, D., Borelli, T., & Mattei, F. (2013). *Diversifying Food and Diets*. London: Routledge.
19. Giuliani, A., Hintermann, F., Rojas, W., & Padulosi, S. (2012). Biodiversity of Andean grains: balancing market potential and sustainable livelihoods. *Bern University of Applied Sciences*. ISBN 978-92-9043-932-5.
20. Glenn, K. C., Alsop, B., Bell, E., Goley, M., Jenkinson, J., Liu, B., … Vicini, J. L. (2017). Bringing new plant varieties to market: plant breeding and selection practices advance beneficial characteristics while minimizing unintended changes. *Crop Science*, 57(6), 2906–2921.
21. Gödecke, T., Stein, A. J., & Qaim, M. (2018). The global burden of chronic and hidden hunger: trends and determinants. *Global Food Security*, 17, 21–29.
22. Gruber, K. (2017). Agrobiodiversity: the living library. *Nature*, 544(7651), S8–S10.
23. Hajjar, R., Jarvis, D. I., & Gemmill-Herren, B. (2008). The utility of crop genetic diversity in maintaining ecosystem services. *Agriculture, Ecosystems & Environment*, 123(4), 261–270.
24. Sangija, F., Martin, H., & Matemu, A. (2021). African nightshades (Solanum nigrum complex): The potential contribution to human nutrition and livelihoods in sub-Saharan Africa. Comprehensive Reviews in Food Science and Food Safety, 20(4), 3284–3318.
25. Hughes, A. R., & Stachowicz, J. J. (2011). Seagrass genotypic diversity increases disturbance response via complementarity and dominance. *Journal of Ecology*, 99, 445–453. doi: https://doi.org/10.1111/j.1365-2745.2010.01767.x
26. Hughes, J. D. A., & Ebert, A. W. (2011). Research and development of underutilized plant species: the role of vegetables in assuring food and nutritional security. In *II International Symposium on Underutilized Plant Species: Crops for the Future-Beyond Food Security* 979 (pp. 79–92), Kuala Lumpur (Malaysia).
27. Ismail, A., Tiong, N. W., Tan, S. T., & Azlan, A. (2009). Antioxidant properties of selected non-leafy vegetables. *Nutrition & Food Science*, 39(2), 176–180.
28. Jaiswal, V., Chauhan, S., & Lee, H. J. (2021). The bioactivity and phytochemicals of Pachyrhizus erosus (L.) Urb.: a multifunctional underutilized crop plant. *Antioxidants*, 11(1), 58.
29. Kaur, J., Rawat, A., Renu, S. K., & Narain, S. (2016). Taxonomy, phytochemistry, traditional uses and cultivation of Ipomoea Aquatica Forsk. *Imperial Journal of Interdisciplinary Research*, 2(10), 408–412.
30. Khan, F., Chai, H. H., Ajmera, I., Hodgman, C., Mayes, S., & Lu, C. (2017). A transcriptomic comparison of two Bambara groundnut landraces under dehydration stress. *Genes*, 8(4), 121.
31. Khoury, C. K., Bjorkman, A. D., Dempewolf, H., Ramirez-Villegas, J., Guarino, L., Jarvis, A., … Struik, P. C. (2014). Increasing homogeneity in global food supplies and the implications for food security. *Proceedings of the National Academy of Sciences*, 111(11), 4001–4006.

32. Kumar, A., Tomer, V., Kaur, A., Kumar, V., & Gupta, K. (2018). Millets: a solution to agrarian and nutritional challenges. *Agriculture & Food Security*, 7(1), 1–15.
33. Li, X., & Siddique, K. H. M. (2018). *Future Smart Food—Rediscovering Hidden Treasures of Neglected and Underutilized Species for Zero Hunger in Asia*. Bangkok: FAO.
34. Li, X., Yadav, R., & Siddique, K. H. (2020). Neglected and underutilized crop species: the key to improving dietary diversity and fighting hunger and malnutrition in Asia and the Pacific. *Frontiers in Nutrition*, 7, 593711.
35. Mabhaudhi, T., Chimonyo, V. G. P., Hlahla, S., Massawe, F., Mayes, S., Nhamo, L., & Modi, A. T. (2019). Prospects of orphan crops in climate change. *Planta*, 250(3), 695–708.
36. Mayes, S., Massawe, F. J., Alderson, P. G., Roberts, J. A., Azam-Ali, S. N., & Hermann, M. (2012). The potential for underutilized crops to improve security of food production. *Journal of Experimental Botany*, 63(3), 1075–1079.
37. Mazahib, A. M., Nuha, M. O., Salawa, I. S., & Babiker, E. E. (2013). Some nutritional attributes of Bambara groundnut as influenced by domestic processing. *International Food Research Journal*, 20, 1165–1171.
38. Migliozzi, M., Thavarajah, D., Thavarajah, P., & Smith, P. (2015). Lentil and kale: Complementary nutrient-rich whole food sources to combat micronutrient and calorie malnutrition. *Nutrients*, 7(11), 9285–9298.
39. Murevanhema, Y. Y., & Jideani, V. A. (2013). Potential of Bambara groundnut (Vigna subterranea (L.) Verdc) milk as a probiotic beverage—a review. *Critical Reviews in Food Science and Nutrition*, 53(9), 954–967.
40. Ngameni, B., Fotso, G. W., Kamga, J., Ambassa, P., Abdou, T., Fankam, A. G., … Kuete, V. (2013). Flavonoids and related compounds from the medicinal plants of Africa. In *Medicinal Plant Research in Africa* (pp. 301–350). Elsevier. http://dx.doi.org/10.1016/B978-0-12-405927-6.00009-6
41. Ngigi, M. W., Okello, J. J., Lagerkvist, C. L., Karanja, N. K., & Mburu, J. (2011). Urban consumers' willingness to pay for quality of leafy vegetables along the value chain: the case of Nairobi Kale consumers, Kenya. International Journal of Business and Social Science, 2(7), 208–216.
42. Nkumah, O. C. (2015). Phytochemical analysis and medicinal uses of Hibiscus sabdariffa. *International Journal of Herbal medicine*, 2(6), 16–19.
43. Nubé, M., & Voortman, R. L. (2011). Human micronutrient deficiencies: linkages with micronutrient deficiencies in soils, crops and animal nutrition. *Combating Micronutrient Deficiencies: Food-Based Approaches*, 7, 289.
44. Ojiewo, C. O., Tenkouano, A., & Yang, R. (2010). The role of AVRDC-The world vegetable centre in vegetable value chains. *African Journal of Horticultural Science*, 3, 1–23.
45. Oyeyinka, S. A., Tijani, T. S., Oyeyinka, A. T., Arise, A. K., Balogun, M. A., Kolawole, F. L., … Joseph, J. K. (2018). Value added snacks produced from Bambara groundnut (Vigna subterranea) paste or flour. *LWT*, 88, 126–131.
46. Pingali, P. L. (2012). Green revolution: impacts, limits, and the path ahead. *Proceedings of the National Academy of Sciences*, 109, 12302–12308. doi: 10.1073/pnas.0912953109
47. Powderly, W. G. (2019). How infection shaped history: lessons from the Irish famine. *Transactions of the American Clinical and Climatological Association*, 130, 127–135.
48. Primiani, C. N., Lestari, U., Amin, M., & Sumitro, S. B. (2013). The dynamics of daidzein chemical compound from yam tuber (Pachyrhizus erosus) tuber and its potential in myometrium. *Berkala Penelitian Hayati*, 18(2), 122–128.
49. Samec, D., Pavlović, I., & Salopek-Sondi, B. (2017). White cabbage (Brassica oleracea var. capitata f. alba): botanical, phytochemical and pharmacological overview. Phytochemistry Reviews, 16(1), 117–135.

50. Sathe, S. K. (1996). The nutritional value of selected Asiatic pulses: chickpea, black gram, mung bean and pigeon pea. In: E. Nwokolo, J. Smartt (eds), *Food and Feed from Legumes and Oilseeds* (pp. 12–32). Boston, MA: Springer.
51. Shakkarpude, J., Jain, A., & Bhimte, A. (2020). Active ingredients of herbal drugs. *Journal of Pharmaceutical and Biological Sciences*, 8(2), 92.
52. Shankul, K. (2013). Systematic pharmacognostical, phytochemical and pharmacological review on an ethno medicinal plant, Basella alba L. *Journal of Pharmacognosy and Phytotherapy*, *5*(4), 53–58.
53. Solanke, S. B., & Tawar, M. G. (2019). Phytochemical information and pharmacological activities of eggplant (Solanum Melongena L.): a comprehensive review. *EAS Journal of Pharmacy and Pharmacology*, 1(5), 106–114.
54. Song, Y. O. (2004). The functional properties of kimchi for the health benefits. *Food Industry and Nutrition*, *9*(3), 27–33.
55. Takahashi, Y., Somta, P., Muto, C., Iseki, K., Naito, K., Pandiyan, M., … Tomooka, N. (2016). Novel genetic resources in the genus vigna unveiled from gene bank accessions. *PLoS ONE*, 11, e0147568.
56. Tan, S. P., Kha, T. C., Parks, S. E., & Roach, P. D. (2016). Bitter melon (Momordica charantia L.) bioactive composition and health benefits: a review. *Food Reviews International*, 32(2), 181–202.
57. United Nations General Assembly. (2015). *Transforming Our World: the 2030 Agenda for Sustainable Development*. New York: United Nations.
58. Wang, T. Y., Li, Q., & Bi, K. S. (2018). Bioactive flavonoids in medicinal plants: structure, activity and biological fate. *Asian Journal of Pharmaceutical Sciences*, 13(1), 12–23.
59. Weinberger, K. (2006, December). Are indigenous vegetables underutilized crops? Some evidence from Eastern Africa and South East Asia. In *International Conference on Indigenous Vegetables and Legumes. Prospectus for Fighting Poverty, Hunger and Malnutrition* 752 (pp. 29–34).
60. Wijesuriya, A. P., Pallemulla, S., Thavarajah, D., Thavarajah, P., & Vandenberg, A. (2013). *Nutritional Impact of Iron Rich Lentil Diet on Iron Deficient Anemic Children in Sri Lanka*. Final Project Report. Colombo: Lady Ridgeway Hospital.
61. World Health Organization. (2021). The State of Food Security and Nutrition in the World 2021: Transforming food systems for food security, improved nutrition and affordable healthy diets for all (Vol. 2021). Food & Agriculture Org. https://www.sidalc.net/search/Record/dig-fao-it-20.500.14283-cb4474en/Description
62. Yulvianti, M., & Zidorn, C. (2021). Chemical diversity of plant cyanogenic glycosides: an overview of reported natural products. *Molecules*, 26(3), 719.

2 Neglected and Underutilized Medicinal Plants of Northwestern Himalayas

Reetika Mahajan
University of Toledo
Toledo, Ohio

Mohamad Maqbool Pakhtoon
Sher-e-Kashmir University of Agricultural Sciences and Technology of Kashmir (SKUAST-K)
Srinagar, India
Rabindranath Tagore University
Patna, India

Faheem Shehjar
University of Toledo
Toledo, Ohio

Sajad Majeed Zargar
Sher-e-Kashmir University of Agricultural Sciences and Technology of Kashmir (SKUAST-K)
Srinagar, India

Zahoor A. Shah
University of Toledo
Toledo, Ohio

2.1 INTRODUCTION

India is one of the important hotspots for diverse species, including plant and animals. According to a report, India covers about 2.4% world area and accounts for 7% of the world's total plant species [1]. Plants have been used as a source of medicine for both humans and animals for many hundreds of years [2]. Due to poverty and resource scarcity, sustainable food and health are currently top priorities in overpopulated developing nations like India. Many plants and the portions of them have been used by hilly

DOI: 10.1201/9781003477792-2

and indigenous people as sources of alternative food and medicine that are typically not consumed by the general public in civilized societies [2]. Underutilized species are those to which agricultural researchers, plant breeders and politicians pay little to no attention. In India, as most of the population lives in rural area, traditional medicine is in practice, including Ayurveda, Yoga, Unani, Siddha, and Homeopathy (AYUSH). Most of the medicinal plants are cultivated in some of the states from North India (Uttar Pradesh, Rajasthan, Himachal Pradesh, Uttarakhand, and Jammu and Kashmir); East India (Bengal, Bihar, Assam); West India (Gujarat, Madhya Pradesh, Maharashtra); South India (Tamil Nadu, Karnataka) [3]. There are about 48,655 documented plant species present in India out of which 7,500 are medicinal plants [4,5]. These identified plant species are mostly found in three major hotspots of biodiversity, i.e. Himalayan, Western Ghats and Andaman and Nicobar Islands. The diverse demography and changing climate of alpine region of Indian Himalayas made this region a great source of important medicinal plants. Till date, the majority of Indian rural and remote region population uses medicinal plants as a primary source of healthcare as these medicines provide significant economic benefits. With the increased interest in exploring medicinal values of plant-based secondary metabolites and awareness about alternatives to allopathy medicines, so far 75% herbal drugs have been developed from traditional medicinal plants [6,7]. The majority of the population in developing countries depends on herbal medicines and traditional medicinal practice for primary healthcare. To date, around 122 phytoconstituents discovered from 94 different plant species are in use for curing a number of diseases [7]. Despite having economic importance, the benefits of most of the medicinal plants are not widely known until now. Moreover, current loss of biodiversity and unawareness of traditional knowledge is becoming a huge threat to medicinal plants and their associated knowledge. Increased commercialization of medicinal plant resources also brings in issues of access to this wealth and equitable sharing of benefits.

2.2 STATUS OF NEGLECTED MEDICINAL PLANTS IN INDIA

India is called as 'Medicinal Garden' of the world because this country has been blessed with most of the important and neglected medicinal plants [8]. According to the biological resources survey of India, Western Ghats is at fifth position throughout the world with about 4,000 medicinal plants identified, whereas the Himalayan region of India has an estimate of 1,748 highly endemic medicinal plant species [9,10]. In medicinal plants, a total of 1,748 species belonging to 915 genera and 223 families are found in the Himalayan region of India (IHR). The families that signify the greatest number of medicinal plants in descending orders are, Asteraceae, Fabaceae, Lamiaceae, Rubiaceae, Euphorbiaceae, Ranunculaceae, Rosaceae, Poaceae, Orchidaceae, Polygonaceae, and Gentianaceae. The medicinal plants-rich genera are *Polygonum, Euphorbia, Piper, Ficus, Aconitum and Swertia, Artemisia, Solanum, Berberis, Desmodium, Allium, and Saussurea*. A total of 675 wild edible plants representing 384 genera and 149 families are known in IHR. Of these, 285 species are herbs, 172 species of shrubs, 197 species of trees, 12 species of pteridophytes, 07 species of fungi and 02 species of lichens. Few significant underutilized medicinal plants from Indian Western Ghats include *Strychnos nuxvomica* (nux vomica), *Abrus precatorius* (crab's eye), *Leptadenia reticulata* (jivanti), *Terminalia chebula* (black myrobalan), *Nothapodytes nimmoniana* (ghanera), *Aristolochia indica* (Indian birthwort) and *Achyranthes aspera* (prickly chaff) [11]. Due to various environmental factors and human activities, these medicinal plants are dwindling very fast throughout the world. Thus, it is important to revive the traditional medicinal system. In India, most threatened medicinal plants are found in Jammu and Kashmir, followed by Sikkim, Himachal Pradesh, Arunachal Pradesh and Uttarakhand. Some of the important underutilized/ neglected medicinal plants of India are given in Table 2.1.

TABLE 2.1
List of Important Underutilized and Neglected Medicinal Plants of India

S. No	Scientific Name and Common Name	Geographical Location	Importance	References
1	*Alternanthera sessilis* (sessile joyweed, dwarf copperleaf)	Assam	Used in treating fever, diarrhea, anemia, liver diseases	[12]
2	*Sesbania grandiflora* (vegetable humming bird, West Indian pea, Agathi or katurai)	Assam, Maharashtra, Tamil Nadu, Uttar Pradesh	Used for headache, fever, congestion, sore throat	[13]
3	*Nymphaea* spp. (water lily, pond lily)	Meghalaya	Remedy for diabetes, liver and urinary disorders, menstruation problem	[14]
4	*Portulaca oleracea* (Purslane)	Throughout India ascending Himalayas	Rich in β carotene, folic acid, vitamin C and essential fatty acids, used in the treatment of osteoporosis and psoriasis	[15]
5	*Talinum triangulare* (water leaf)	Karnataka, Tamil Nadu, Kerala	Used as diuretic, in gastrointestinal disorders, high blood pressure, anemia	[16]
6	*Ipomoea aquatica* (water spinach)	Kerala and Tamil Nadu	Against piles, nosebleeds and high blood pressure	[17]
7	*Sauropus androgynus* (Chekurmanis)	Sikkim, Himalayas, Khasi, Abor and Aka hills and in the Western Ghats of Kerala	High nutritive value, rich in protein, minerals and vitamin A, B, C and antioxidants	[18]
8	*Solanum torvum* (turkey berry, wild eggplant)	Tamil Nadu, Tripura, Northeast India	Treatment of fever, wounds, tooth decay, reproductive problems and hypertension	[19]
9	*Vigna umbellata* (rice bean, climbing bean, mountain bean)	Northeastern hills and the Western and Eastern Ghats	Multipurpose legume eaten as vegetable, rich in vitamins and proteins	[20]

(*Continued*)

TABLE 2.1 (*Continued*)
List of Important Underutilized and Neglected Medicinal Plants of India

S. No	Scientific Name and Common Name	Geographical Location	Importance	References
10	*Dendrocalamus strictus* (male bamboo, solid bamboo, and calcutta bamboo)	Eastern and northeast part of India	Therapy for cold, cough and fever	[21]
11	*Coptis teeta* (Mamira)	Arunachal Pradesh	Antitumor and antibacterial activities	[22]
12	*Gymnocladus assamicus* (coffee tree)	Arunachal Pradesh	Used for treating swelling and abscesses	[23]
13	*Illicium griffithii* (star anise and anise tree)	Northeastern states of India	An antifungal agent and food preservative	[24]
14	*Lilium polyphyllum* (white Himalayan lily)	Northwestern Himalayan region	Used for treating bronchial asthma and cough	[25]
15	*Nardostachys jatamansi* (muskroot)	Eastern Himalayans	For epilepsy and mental weakness	[26]

2.3 NORTHWESTERN HIMALAYAS

The Himalayan Mountain Range straddles a transition zone between the Palearctic and Indo-Malayan realms. The Indian Himalayan region, which includes the parts of Trans, North-west, West, Central, and East Himalaya and ranges from Arunachal Pradesh to Jammu and Kashmir and Ladakh Union Territories, is located in between 27°–38° N latitudes and 72°–89° E longitudes and an altitude ranging from 200 to >8,000 m above mean sea level covering around 18% of India, i.e., 419,873 km^2 (~2,800 km long and 220–300 km wide). Northwestern Himalayas includes three important states of India (Jammu and Kashmir, Himachal Pradesh and Uttarakhand). The Indian Himalayan region is famous for one of the important biodiversity hotspots of the world. In addition, to the geological, climatic and altitudinal variations in the region, it is also known for its representative, natural, unique and socio-economically important biodiversity as well as topographic complexity which again contributes to the biological diversity of the mountains along their east-west and north-south axes [27,28]. The rich plant diversity has been utilized as medicine, food (edible), fuel, fodder and for making agricultural tools, house building, small-scale enterprises such as making basket, mat, hat, and kilta by the locals [12].

The database on plant species from the Indian Himalayan region, Nepal and Bhutan revealed that there are about 10,452 species of angiosperms which belong to 2,302 genera and 232 families and 51 species of gymnosperms which belong to 20 genera and eight families [29]. A total of 18,940 species of plants present in the Indian Himalayan region are enlisted as endemic.

In angiosperms overall, 6,745 species of 225 families, belonging to 1,768 genera, have been noticed from the Trans, Northwest, and Western Himalayas. Among that,

TABLE 2.2
Two Major Plant Groups of the Indian Himalayan Region

Plant Group	Different Families of Plants	Reference
Gymnosperms	Cupressaceae, Cycadaceae, Ephedraceae, Ginkgoaceae, Pinaceae, Podocarpaceae, Taxaceae, and Taxodiaceae	[29]
Angiosperms	Acanthaceae, Apiaceae, Asteraceae, Brassicaceae, Fabaceae, Euphorbiaceae, Lamiaceae, Liliaceae, Orchidaceae, Poaceae, Polygonaceae, Rosaceae, Rubiaceae, Scrophulariaceae, etc.	[12]
	Monotypic families representing single species	
	Actinidiaceae, Adoxaceae, Casuarinaceae, Biebersteiniaceae, Calycanthaceae, Cannaceae, Caricaceae, Ceratophyllaceae, Circaesteraceae, Clusiaceae, Coriariaceae, Daphniphyllaceae, Datiscaceae, Dipterocarpaceae, Droseraceae, Hippuridaceae, Icacinaceae, Iteaceae, Lardizabalaceae, Leeaceae, Marantaceae, Martyniaceae, Melianthaceae, Moringaceae, Myricaceae, Ochnaceae, Paeoniaceae, Platanaceae, Plumbaginaceae, Podophyllaceae, Podostemaceae, Proteaceae, Punicaceae, Rhizophoraceae, Saurauiaceae, Sonneratiaceae, Saururaceae, Sphenocleaceae, Stylidiaceae, Torricelliaceae, Trillidiaceae, Tropaeolaceae, Zannichelliaceae.	

604 were tree species, 1,049 shrubs and herbs are 5,092 [12]. About 3984 species of angiosperms and 21 species of gymnosperms are reported from Arunachal Pradesh; 3,010 species of angiosperms and seven species of gymnosperms from Assam; 2,376 species of angiosperms and five species of gymnosperms from Manipur; 1,886 species of angiosperms and six species of gymnosperms from Meghalaya; 2,141 species of angiosperms and six species of gymnosperms from Mizoram; 2,431 species of angiosperms and nine species of gymnosperms from Nagaland; 1,463 species of angiosperms and 13 species of gymnosperms from Tripura; and 4,458 species of angiosperms from Sikkim were recorded [30]. The dominating genera in Trans, Northwest, and West Himalaya are Carex, Taraxacum, Potentilla, Astragalus, Saxifraga, Cotoneaster, Artemisia, Cyperus, Polygonum, Corydalis, Berberis, Euphorbia, Silene, Poa, Primula, Pedicularis, Nepeta, Impatiens, Ranunculus, Persicaria, Veronica, Allium, Rubus, Ficus, etc. (26). The plant species found in Indian Himalayan region is classified into two major groups: angiosperm and gymnosperm (Table 2.2).

2.4 UNDERUTILIZED MEDICINAL PLANTS OF JAMMU AND KASHMIR

Traditional medicine relies on medicinal plants since they are cheaper, more efficient, and have a natural origin. Millions of people in Asian countries are treated by traditional doctors using herbs [31]. With its wide range of altitudes, topography, and climatic conditions, the northwestern Himalayan region, especially Jammu and

Kashmir, is a rich source of medical wealth and has a significant role in Vedic literature [32]. The traditional methods of treating various ailments with diverse medicinal plants were the foundation of the Asian people's understanding and practice of animal health. These therapeutic herbs, though neglected in conventional medicine, are increasingly accepted in the preparation of ethnomedicine [33]. The underutilized plants with high medicinal potential in Jammu and Kashmir that were discussed include *Nardostachys jatamansi*, *Portulaca oleracea*, *Saussurea costus*, *Fagopyrum esculentum*, *Heracleum candicans*, *Glycyrrhiza glabra*, *Rumex acetosa*, etc. The roots, leaves, flowers, stems, and gum are used as medicine to treat a variety of illnesses such as painful urination, epilepsy, convulsions, etc. [31].

It has been discovered that using medicinal plants in traditional medicine is efficient, affordable, and useful. Due to the difficult economic climate, the high cost of pharmaceuticals, and their limited availability, the practice is rapidly expanding [33]. Skin conditions, lung infections, diarrhea, fever, eye infections, wounds, external and internal parasites, poor milk secretion, inadequate weight gain, loss of appetite and exhaustion are just a few of the disorders that can be treated using these underutilized medicinal plants. The leaves, barks, tubers, roots, herbs and plant extracts are only a few of the parts of medicinal plants that are utilized for various therapies. Alkaloids, terpenes, phenolic compounds, basic metabolites, glycosides and secondary metabolites are among the substances that these sections secrete [33,34]. Decoction, emulsion, apozems, liniments, electro activities and powdered medicinal plant preparations are just a few of the many forms that they can take. Equal use of these underutilized medicinal plants is made in the food, cosmetic and fragrance sectors. There is a chance to find the evolution of pharmaceuticals in neglected medicinal plants because the pharmaceutical industry first extracts the active compounds before using them to manufacture drugs. Common food shortages are reduced by the alternate food sources provided by underutilized plant species. The study of novel staple crops and the promotion of the extraction, utilization and conservation of underutilized species spur sustainable consumption [35]. Some of the known underutilized medicinal plants of Jammu and Kashmir are listed here Table 2.3.

TABLE 2.3
List of the Underutilized Medicinal Plants in Jammu and Kashmir

S No	Local Name	Scientific Name	Medicinal Value	References
01	Spikenard	*Nardostachys jatamansi*	Ayurvedic classics	[26]
02	Nunner	*Portulaca oleracea*	Anticancerous potential	[36]
03	Putchuk	*Saussurea costus*	Antiseptic, bronchial asthma, leprosy, insecticide antiulcerogenic	[37]
04	Buckwheat	*Fagopyrum esculentum*	antioxidant potential	[38]
05	Hogweed	*Heracleum candicans*	Medicinal herb	

(*Continued*)

TABLE 2.3 (*Continued*)
List of the Underutilized Medicinal Plants in Jammu and Kashmir

S No	Local Name	Scientific Name	Medicinal Value	References
06	Licorice	*Glycyrrhiza glabra*	Anti-tuberculosis potential	[39]
07	Abej	*Rumex acetosa*	Medicinal herb and phytoremediation	[40]
08	Ulam Gunung	*Hydrocotyle sibthorpioides* Lam.	Fever, edema, detoxication, antidiuretic activity, throat pain and antitumor	[41]
09	Kandara	*Cynara scolymus* L.	Hepatoprotective, lowers cholesterol level, anti-diabetic	[41]
10	Stalkless Joyweed	*Alternanthera sessilis*	Diarrhea, dysentery, fevers, eye drop	[37]
11	Oregano	*Origanum vulgare*	Anti-cancer, antioxidant, Anti-inflammatory	[42]
12	Chora, Choru, Gandrayan	*Angelica glauca Edgew*	Bronchitis, constipation	[37]
13	Skotche	*Allium przewalskianum*	Helps reduce blood cholesterol levels, acts as a tonic to the digestive system and also tonifies the circulatory system	
14	Nodding Beggarticks	*Bidens cernua* L.	Urinary tract infections	[37]
15	Pipo fruit	*Solanum muricatum*	Anti-cancer, diuretic, kidney asher and improves immune system function	[43]
16	Gunja	*Abrus precatorius* L	To treat tetanus and to prevent rabies.	[4]
17	Scarlet pimpernel	*Anagallis arvensis*	Anti-mycotic, antimicrobial, molluscicidal, antioxidant, anti-inflammatory, anti-leishmania, antiviral, cytotoxic and spermatogenesis.	[44]
18	Coriander	*Coriandrum sativum*	Stomach and intestinal problems, including upset stomach, nausea, diarrhea, intestinal gas, constipation and irritable bowel syndrome (IBS).	[45]
19	Lavender	*Lavandula angustifolia Mill.*	Antibacterial and antifungal effects.	[46]
20	Wild Rue	*Peganum harmala*	Relieving remittent and chronic fevers.	[46]

2.5 DESCRIPTION AND USES OF THE MEDICINAL PLANTS

2.5.1 *Rumex acetosella*

It is a rhizome-spreading perennial herb. It has a straight, slender, reddish stem that branches off at the top and grows to a height of 50 cm (20 in). The tiny, arrow-shaped leaves have sharp lobes at the base and measure only 2.4–5 cm (1–2 in) long by. 5–2 cm wide. It blossoms from March to November when the male or female plants' apex-located yellowish-green (male) or reddish (female) flowers begin to ripen into the bright red fruits.

2.5.2 *Nardostachys jatamansi*

A flowering member of the honeysuckle family, jatamansi is mostly found in Jammu and Kashmir in northwestern Himalayas. The plant produces pink bell-shaped flowers and can reach a height of 10–50 cm (4–20 in).

2.5.3 *Portulaca oleracea*

It develops into a spot that is centered on the main root and is a prostrate plant that is sometimes tufted, heavily branched from the base. It may expand to a diameter of 60 cm. The branches may produce a taproot or other roots. The stem is cylindric, substantial, lush, and full.

2.5.4 *Cynara scolymus* L.

The artichoke is a hardy perennial plant that can reach a height of 2 m. It has broad, flat, or pinnatifid leaves that run-in arch to sharp thorns. The majority of developed varieties today lack spines. The large flower heads are 8–15 cm in diameter. The fleshy receptacles are ringed by blunt husk leaves that are arranged like roof tiles and are filled with blue-violet tubular blooms.

2.5.5 *Abrus precatorius*

It is a herbaceous blooming plant in the bean family Fabaceae, also called the rosary pea or the jequirity bean. It is a tall, pinnately leafed perennial climber that wraps itself around hedges, bushes and trees. The plant is most well-known for its seeds, which are toxic due to the presence of abrin and are used as beads and in percussion instruments. One seed, thoroughly chewed, can be lethal to both adults and kids. The plant is indigenous to Australia and Asia. Where it has been introduced, it has a propensity to spread like a weed and become invasive.

2.5.6 *Lavandula angustifolia* Mill

It is a highly fragrant shrub that can reach heights of up to two meters (3 ft, 3 in–6 ft, 7 in). The leaves are 2–6 cm (1–2 + 12 in) long, 4–6 mm ($^{3}/_{16}$–$^{1}/_{4}$ in) wide and evergreen. At the apex of slender, leafless stems 10–30 cm (4–12 in) long, spikes 2–8 cm (1–3 in) long bearing pinkish-purple (lavender-colored) flowers are formed.

2.5.7 *Allium przewalskianum*

It features tiny bulbs that can be up to 10 mm wide and up to 40 cm tall and round in cross-section, scape. The tubular leaves are approximately the same length as the scape. It helps in reducing blood cholesterol levels, acts as a tonic to the digestive system, and also tonifies the circulatory system.

2.5.8 *Solanum muricatum*

It is a 4–6 ft tall little bush or shrub, similar to the tomato. The leaves are roughly 15 cm long, green, and elongated. The tiny, white and purple flowers grow in groups. From May to September, flowers bloom. It is anti-cancer, diuretic, kidney asher and improves immune system function.

2.6 UNDERUTILIZED MEDICINAL PLANT OF HIMACHAL PRADESH AND UTTARAKHAND

Himachal Pradesh and Uttarakhand are other two biodiversity hotspots of the north-western Himalayan region of India. Both these states are hilly states of India which is a habitat to various valuable medicinal plants [47]. Some of the important under-utilized medicinal plants of these two regions are enlisted in Table 2.4.

TABLE 2.4
List of Various Underutilized/Neglected Plants with Medicinal Values of Himachal Pradesh and Uttarakhand

S.No	Scientific Name	Local Name	Uses	References
1	*Rubus ellipticus* Sm	Yellow Himalayan raspberry, Hisalu, Ashilo	Provides free energy packets for the people traveling to mountains	[48]
2	*Fragaria × ananassa*	Strawberry	Great source of iron, anti-diabetic	[49]
3	*Prunus persica* (L.) Batsch	Plum	Wound healing	[50]
4	*Myrica esculenta* Buch.-Ham. ex D. Don	Kafal or Kaphal	Beneficial in cardiac debility, edema and haemoptysis and have ulcer healing properties	[51]
5	*Ficus palmata* Forssk.	Khemri	Beneficial in the disease of lungs and the bladder	[52]
6	*Punica granatum* L.	Darim	Traditional remedy against diarrhea, dysentery and intestinal parasites	[53]
7	*Berberis asiatica* Roxb. ex DC.	Chutro, Marpyashi, Daruharidra, Darbi	Antimicrobial, wound healing, hepatoprotective and cytotoxicity etc	[54]
8	*Solanum nigrum* L.	Black Nightshade	Expectorant, analgesic, sedative and diaphoretic properties. Its external application cures skin diseases and gives relief in burns, itching, pain etc	[55]

(*Continued*)

TABLE 2.4 (*Continued*)
List of Various Underutilized/Neglected Plants with Medicinal Values of Himachal Pradesh and Uttarakhand

S.No	Scientific Name	Local Name	Uses	References
9	*Pyracantha crenulata* (Roxb. ex D.Don) M.Roem	Ghigharu	Cardio-tonic, coronary vasodilator and hypertensive properties	[56]
10	*Prunus armeniaca* L	Khubani	Alternative treatment for cancer	[57]
11	*Rosa canina* L.	Dog Rose	Treatment of viral infections and disorders of the kidneys and urinary tract	[58]
12	*Rubus occidentalis* L.	Black raspberry	Beneficial for cancer treatment	[59]
13	*Ficus auriculata* Lour.	Timul, Timil	Treating gastrointestinal problems; have hypoglycemic, anthelmintic, anti-tumor and hypotensive activity	[60]
14	*Ziziphus jujube*	Ber	Antidote, diuretic, emollient and expectorant, promotes hair growth. anodyne, anti-cancer, pectoral, refrigerant, blood purifier and aid in digestion. treatment of chronic fatigue, loss of appetite, diarrhea, anemia, irritability and hysteria, hypnotic, narcotic, sedative, stomachache, treatment of palpitations, insomnia, nervous exhaustion, night sweats and excessive perspiration. Dyspepsia	[61]
15	*Vitis vulpina* L.	Forest grape	Treating sore throats, and raisins were given as treatments for consumption (tuberculosis), constipation, cancer, cholera, smallpox, nausea, skin and eye infections as well as kidney and liver diseases.	[61]
16	*Ribes nigrum* L.	Blackcurrant	Treating rheumatoid arthritis and night and fatigue-related visual impairment, antimicrobial and anti-cancer properties	[62]
17	*Ficus carica* L.	Indian fig, Edible Fig	Colic, indigestion, diarrhea, sore throats, coughs, bronchial problems, inflammatory, cardiovascular disorders, ulcerative diseases and cancers	[63]

(*Continued*)

TABLE 2.4 (*Continued*)
List of Various Underutilized/Neglected Plants with Medicinal Values of Himachal Pradesh and Uttarakhand

S.No	Scientific Name	Local Name	Uses	References
18	*Syzygium cumini* (L.) Skeels	Jamun, Java Plum, Black Plum, Jambul and Indian blackberry	For curing diarrhea, stomachache, carminative and diuretic and have cooling and digestive properties	[64]
19	*Cydonia oblonga* Mill.	Bihi	Sedative, antipyretic, anti-diarrheic and antitussive properties and for the treatment of various skin diseases	[65]

2.7 CONCLUSION AND FUTURE PERSPECTIVE

In India, 21.71% of total geographical area is covered by forests which makes it one of the most important biodiversity hotspots in the world. Most of the Indian population is living in rural area where everyone is dependent on the traditional medicine system for curing a disease. These people are healthier than the urban area population as knowingly or unknowingly they consume various underutilized or neglected medicinal plants in their day-to-day life. There are many incidents in India where one discovers medicinal potential of a plant accidently. Despite having diverse medicinal plants, not much is known by the scientific community till date. Thus, there is a need to explore more and more underutilized/neglected plants with medicinal value so that these plants can be commercialized and cultivated by locals to earn a livelihood. Government and Non-Government Organizations (NGOs) should start awareness camps with the help of scientists and locals to educate the masses related to the conservation and cultivation of these neglected and underutilized medicinal plants.

REFERENCES

1. Chitale VS, Behera MD, Roy PS (2014) Future of endemic flora of biodiversity hotspots in India. *PLoS ONE* 9(12):e115264. https://doi.org/10.1371/journal.pone.0115264
2. Omotayo AO, Aremu AO (2020) Underutilized African indigenous fruit trees and food–nutrition security: opportunities, challenges, and prospects. *Food and Energy Security* 9(3):e220.
3. Ravishankar B, Shukla VJ (2007) Indian systems of medicine: a brief profile. *African Journal of Traditional, Complementary and Alternative Medicines* 4(3):319–337.
4. Kumari S, Batish DR, Singh HP, Negi K, Kohli RK (2013) An ethnobotanical survey of medicinal plants used by Gujjar community of Trikuta Hills in Jammu and Kashmir, India. *Journal of Medicinal Plants Research* 7:2111–2121.
5. Sharma N, Pandey R (2013) Conservation of medicinal plants in Tropics. In: Normah MN, Chin HF, Reed BM (eds), *Conservation of tropical plant species*. Springer, New York, pp 437–487. https://doi.org/10.1007/978-1-4614-3776-5

6. World Health Organization (2011) *The world traditional medicines situation, in traditional medicines: Global situation, issues and challenges*. Geneva, 3, 1–14.
7. Singh G, Rawat GS (2011) Ethnomedicinal survey of Kedarnath wildlife sanctuary in Western Himalaya, India. *Indian Journal of Fundamental and Applied Life Science* 1:35–46.
8. Zakiullah Z, Ahmadullah A, Khisroon M, Saeed M, Khan A, Khuda F,... Ismail M (2015) Genetic susceptibility to oral cancer due to combined effects of GSTT1, GSTM1 and CYP1A1 gene variants in tobacco addicted patients of Pashtun ethnicity of Khyber Pakhtunkhwa province of Pakistan. *Asian Pacific Journal of Cancer Prevention* 16(3):1145–1150.
9. Joshi RK, Satyal P, Setzer WN (2016) Himalayan aromatic medicinal plants: a review of their ethnopharmacology, volatile phytochemistry, and biological activities. *Medicines* 3(1):6. 10.3390/medicines3010006
10. Patwardhan A, Pimputkar M, Mhaskar M, Agarwal P, Barve N,Gunaga R, ... Vasudeva R (2016) Distribution and population status of threatened medicinal tree Saracaasoca (Roxb.) De Wilde from Sahyadri-Konkan ecological corridor. *Current Science* 111(9):1500–1506.
11. Mallikarjuna Gowda AP, Ranjini TN, Peethambar SK, Praneeth YS (2015) Underutilized, yet potential medicinal plants of Western Ghats. In: III International Symposium on Underutilized Plant Species 1241 (pp 15–22). https://www.ishs.org/ishs-article/1241_3
12. Samant SS (2015) Assessment, valuation and conservation prioritization of floristic diversity in trans, North Western and Western Himalaya [D.Sc. thesis]. Uttarakhand: Kumaun University, Nainital.
13. Rana SK, Rawat GS (2017) Database of Himalayan plants based on published floras during a century. *Data* 2:36. https://doi.org/10.3390/data2040036
14. Naithani HB (2020) *Floristic diversity of North East India*. Rain Forest Research Institute (Indian Council of Forestry Research and Education), Jorhat, Assam. P. 182.
15. Singh RP, Chidambara Murthy KN, Jayaprakasha GK (2002) Studies on the antioxidant activity of pomegranate (Punica granatum) peel and seed extracts using in vitro models. *Journal of Agricultural and Food Chemistry* 50(1):81–86.
16. Brown D, Warschauer M (2006) From the university to the elementary classroom: students' experiences in learning to integrate technology in instruction. *Journal of Technology and Teacher Education* 14(3):599–621.
17. Sindhu, R., Binod, P., & Pandey, A. (2016). Biological pretreatment of lignocellulosic biomass–an overview. *Bioresource Technology* 199:76–82.
18. Mittermeier RA, Gil PR, Hoffman M, Pilgrim J, Brooks T, Mittermeier CG, Lamoreux J, da Fonseca GAB. (2004) Hotspots revisited: Earth's biologically richest and most endangered terrestrial ecoregions. Cemex, Mexico. P. 392.
19. Samant SS, Dhar U, Palni LMS (1998) *Medicinal plants of Indian Himalaya: diversity distribution potential values*. GyanodayaPrakashan, Nainital, India.
20. Singh H (2021) Building effective blended learning programs. In: *Challenges and opportunities for the global implementation of e-learning frameworks* Editors: Badrul H. Khan, Saida Affouneh, Soheil Hussein Salha, Zuheir Najee Khlaif. IGI Global, Pennsylvania pp. 15–23.
21. Vamanu E (2011) Antioxidant and antimicrobial activities of ethanol extracts of Cynara scolymus (Cynarae folium, Asteraceae family). *Tropical Journal of Pharmaceutical Research* 10(6):777–783.
22. Bajpay A, Nainwal RC, Singh D (2019) Coptis teeta: a potential endemic and endangered medicinal plant of Eastern Himalayas. *Journal of Pharmacognosy and Phytochemistry* 8(4):245–248.

23. Choudhury BI, Khan ML, Dayanandan S (2014) Functional androdioecy in critically endangered *Gymnocladus assamicus* (Leguminosae) in the eastern Himalayan region of northeast India. *PLoS ONE* 9(2):e87287. https://doi.org/10.1371/journal.pone.0087287
24. Saraswathy A, Shakila R, Lavanya SM, Arunmozhidevi A (2010) Essential oil constituents of Illicium griffithii and its antimicrobial activity. *Pharmacognosy Magazine* 6(23):208. https://doi.org/10.4103/0973-1296.66938
25. Awotedu BF, Akala AO, Omolola TO, Owoeye EA, Olaoti-Laaro SO, Ogunsiji AO (2021). Some underutilized indigenous timber trees with high medicinal values: its challenges and prospects. *World Journal of Advanced Research and Reviews* 10(2):161–168.
26. Dhiman N, Bhattacharya A (2020) *Nardostachys jatamansi* (D. Don) DC.-Challenges and opportunities of harnessing the untapped medicinal plant from the Himalayas. *Journal of Ethnopharmacology* 246:112211.
27. Gandhi AD, Vizhi DK, Lavanya K, Kalpana VN, Rajeswari VD, Babujanarthanam R. (2017). In vitro anti-biofilm and anti-bacterial activity of Sesbania grandiflora extract against *Staphylococcus aureus. Biochemistry and Biophysics Reports* 12:193–197. https://doi.org/10.1016/j.bbrep.2017.10.004
28. Rahimi VB, Mousavi SH, Haghighi S, Soheili-Far S, Askari VR (2019) Cytotoxicity and apoptogenic properties of the standardized extract of Portulaca oleracea on glioblastoma multiforme cancer cell line (U-87): a mechanistic study. *EXCLI* Journal 18:165. https://doi.org/10.17179/excli2019-1063
29. Rana S, Rawat J, Misra RDK (2005) Anti-microbial active composite nanoparticles with magnetic core and photocatalytic shell: TiO_2–$NiFe_2O_4$ biomaterial system. Acta Biomaterialia 1(6):691–703.
30. Naithani R, Huma LC, Holland LE, Shukla D, McCormick DL, Mehta RG, Moriarty RM (2008) Antiviral activity of phytochemicals: a comprehensive review. Mini Reviews in Medicinal Chemistry 8(11):1106–1133.
31. Delin SHM, Zaimah R (2022) Kebarangkalian Berkahwin Lewat dalam Kalangan Generasi Milenium di Malaysia [The probability of late marriage among the millennial generation in Malaysia]. *Akademika* 92(2):113–127.
32. Lone MA (2014). An interrogation into the relationship of North Western Himalayan tribes and the natural resources. *Online International Interdisciplinary Research Journal* 4(Special Issue (March)):623–632.
33. Belewu MA, Olatunde OA, Giwa TA (2009) Underutilized medicinal plants and spices: chemical composition and phytochemical properties. *Journal of Medicinal Plants Research* 3(12):1099–1103.
34. Evans DA, Brightman CAJ (1980) Pleomorphism and the problem of recrudescent parasitaemia following treatment with salicylhydroxamic acid (SHAM) in African trypanosomiasis. *Transactions of the Royal Society of Tropical Medicine and Hygiene* 74(5):601–604.
35. Kunwar RM, Mahat L, Sharma LN, Shrestha KP, Kominee H, Bussmann RW (2012) Underutilized plant species in far west Nepal. *Journal of Mountain Science* 9(5):589–600.
36. Murtaza I, Rashid N, Laila O, Ganie NA, Wani MY (2020) Anti-cancer and anti-microbial potential of nutraceutical rich underutilized herb portulaca oleracea from Kashmir region. *International Journal of Current Microbiology and Applied Sciences* 9(9):3496–3508.
37. Mehta P, Bisht K, Sekar KC (2021) Diversity of threatened medicinal plants of Indian Himalayan Region. *Plant Biosystems* 155(6):1121–1132.
38. Pirzadah TB, Malik B, Tahir I, Qureshi MI, Rehman RU (2017) Metabolite fingerprinting and antioxidant potential of tartary buckwheat-an underutilized pseudocereal crop from Kashmir Region. *Free Radicals and Antioxidants* 7(1):95–106.

39. Shah A, Rather MA, Hassan QP, Aga MA, Mushtaq S, Shah AM.,... Ahmad Z (2017). Discovery of anti-microbial and anti-tubercular molecules from Fusarium solani: an endophyte of Glycyrrhiza glabra. *Journal of Applied Microbiology* 122(5):1168–1176.
40. Korpelainen H, Pietiläinen M (2020) Sorrel (*Rumex acetosa* L.): not only a weed but a promising vegetable and medicinal plant. *The Botanical Review* 86(3):234–246.
41. Vamanu E (2011) Antioxidant and antimicrobial activities of ethanol extracts of Cynara scolymus (Cynarae folium, Asteraceae family). *Tropical Journal of Pharmaceutical Research* 10(6):777–783.
42. Ashfaqullah S, Tamta S (2018) Studies on high value medicinal plant-origanum vulgare L.: a review.
43. Gorbe M, Bhat R, Aznar E, Sancenón F, Marcos MD, Herraiz FJ, Prohens J, Venkataraman A, Martínez-Máñez R (2016) Rapid biosynthesis of silver nanoparticles using pepino (*Solanum muricatum*) leaf extract and their cytotoxicity on HeLa cells. *Materials* 9(5):325–333.
44. Kumar R, Bhagat N (2012) Ethnomedicinal plants of district Kathua (J&K). *International Journal of Medicinal and Aromatic Plants* 2:603–611.
45. Bhangu A, Ademuyiwa AO, Aguilera ML, Alexander P, Al-Saqqa SW, Borda-Luque G.,... Iftekhar F (2018) Surgical site infection after gastrointestinal surgery in high-income, middle-income, and low-income countries: a prospective, international, multicentre cohort study. *The Lancet Infectious Diseases* 18(5):516–525.
46. Bano H, Siddique MA, Bhat MA (2017) Ethnomedicinal appraisal of medicinal plants used by ethnic communities of district Anantnag and Budgam of Kashmir Himalaya, India. *Plant Archives* 17:1563–1576.
47. Kala CP, Farooquee NA, Dhar U (2004) Prioritization of medicinal plants on the basis of available knowledge, existing practices and use value status in Uttaranchal, India. *Biodiversity & Conservation* 13(2):453–469.
48. Manandhar NE (2002) *Plants and people of Nepal*. Timber Press, Portland, Oregon. P. 599.
49. Ibrahim DS, Abd El Maksoud MA (2015) Effect of strawberry (Fragaria x ananassa) leaves extract on diabetic nephropathy in rats. *International Journal of Experimental Pathology* 96(2):87–93.
50. Hussain T, Baba I, Jain SM (2014) Evaluation of wound healing active principles from prunus persica. *International Journal of Research in Pharmacy and Chemistry* 4(1):233–236.
51. Gusain YS, Khanduri VP (2016) Myrica esculenta wild edible fruit of Indian Himalaya: need a sustainable approach for indigenous utilization. *Ecology, Environment and Conservation* 22:267–270.
52. Bhowmik D, Gopinath H, Kumar BP, Duraivel S, Aravind G, et al. (2013) Medicinal uses of Punica granatum and its health benefits. *Journal of Pharmacognosy and Phytochemistry* 1(5):2278–4136.
53. Komal S, Bairwa R, Chauhan N, Srivastava B (2011) Berberis aristata: a review. *IJRAP* 2(2):383–388.
54. Dehar N, Walia R, Verma RB, Pandey P (2013) hepatoprotective activity of Berberis aristata root extract against chemical induced acute hepatotoxicity in rats. *Asian Journal of Pharmaceutical and Clinical Research* 6(5):53–56.
55. Nyeem MAB, Rashid AKMMU, Nowrose M, Hossain MA (2017) Solanum nigrum (Maku): a review of pharmacological activities and clinical effects. *International Journal of Applied Research* 3(1):12–17.
56. Bisht VK, Kandari LS, Negi JS, Bhandari AK, Sundriyal RC (2013) Traditional use of medicinal plants in district Chamoli, Uttarakhand, India. *Journal of Medicinal Plants Research* 7(15):918–929.

57. Sharma S, Satpathy G, Gupta RK (2014) Nutritional, phytochemical, antioxidant and antimicrobial activity of Prunus armenicus. *Journal of Pharmacognosy and Phytochemistry* 3(3):23–28.
58. Ghrabi Z (2005) *A guide to medicinal plants in North Africa: Rosa canina L.* International Union for Conservation of Nature and Natural Resources, Malaga, Spain, pp. 229–231.
59. Kresty LA, Frankel WL, Hammond CD, Baird ME, Mele JM, Stoner GD, Fromkes JJ (2006) Transitioning from preclinical to clinical chemopreventive assessments of lyophilized black raspberries: interim results show berries modulate markers of oxidative stress in Barrett's esophagus patients. *Nutrition and Cancer* 54(1):148–156.
60. Rout SD, Panda T, Mishra N (2009) Ethnomedicinal plants used to cure different diseases by tribal's of Mayurbhanj district of North Orissa. Ethnomedicine 3(1):27–32.
61. Yu L, Jiang BP, Luo D, Shen XC, Guo S, Duan JA, Tang YP (2012) Bioactive components in the fruits of Ziziphus jujube Mill. against the inflamatory irritant action of Eubhorbia plants. *Phytomedicine* 19(3–4):239–244.
62. Nielsen IL, Haren GR, Magnussen EL, Dragstead LO, Rasmussen SE (2003) Quantification of anthocyanins in commercial black currant juices by simple high-performance liquid chromatography. Investigation of their pH stability and antioxidative potency. *Journal of Agricultural and Food Chemistry* 51(20):5861–5866.
63. Gilani AH, Mehmood MH, Janbaz KH, Khan AU, Saeed SA (2008) Ethnopharmacological studies on antispasmodic and ant platelet activities of Ficus carica. *Journal of Ethnopharmacology* 119(1):1–5.
64. Thaper AR (1958) Jamun, ICAR, Farm Bull 42.
65. Oliveira AP, Pereira JA, Andrade PB, Valentão P, Seabra RM, Silva BM (2007) Phenolic profile of Cydonia oblonga Miller leaves. *Journal of Agricultural and Food Chemistry* 55(19):7926–7930.

3 Neglected and Underutilized Crops with Medicinal Importance

Sabreena, Shahnawaz Hassan, and Bashir Ah Ganai
University of Kashmir, Srinagar, India

3.1 INTRODUCTION

The world population (7.87 billion) is expected to reach 9.6 billion by 2050, growing at a pace of 1.03% each year, and India accounts for 17.5% of the global population, while having just 2.4% of the world's surface area and 1.38 billion people [1]. The biggest problem today is providing consistent, safe, and nutritious food to this constantly rising population [2]. It is necessary to identify native, neglected, resistant/tolerant crops that can withstand climatic changes and acclimatize to a broad assortment of environmental conditions (Figure 3.1). Indigenous crops have shown to be better not just in terms of environmental adaptation but also in terms of nutritional value [3]. Underutilized or minor crops are fantastic alternatives that are historically farmed in the centers of origin or variety by farmers and are less important in global production and consumption systems [4]. They are a key element of the local food and have adapted well to marginal areas. These plants are used in traditional medicine and contain significant nutritional components such as proteins, vitamins, and minerals, making them potentially useful in the treatment of a variety of ailments [4]. For a variety of human requirements, these crops offer important nutritional and industrial value. Micronutrients are plentiful in underutilized crops, including *Panicum miliaceum* (proso millet), *Setaria italica* (Foxtail millet), *Amaranthus* (amaranth), and *Fagopyrum*(buckwheat). While each of these grains may help to establish a more regular diet, they can also be quite helpful in avoiding vitamin deficiencies [5]. Due to favorable weather circumstances, these crops have historically been cultivated in Himachal Pradesh, Uttarakhand, and Jammu & Kashmir [6]. Since the cultivation of these crops is restricted to specific geographic pockets in different agro-ecological regions, primarily by impoverished farming societies who benefit from them and depend on them for their livelihood, the general public is still unaware of their commercial significance, market value, and nutritional and medicinal properties [7]. Due to the presence of different secondary metabolites, these crops are being used extensively for treating oxidative stress-related diseases [8]. Recent scientific evidence shows that, in addition to being a great source of nutrition, due to these secondary

DOI: 10.1201/9781003477792-3

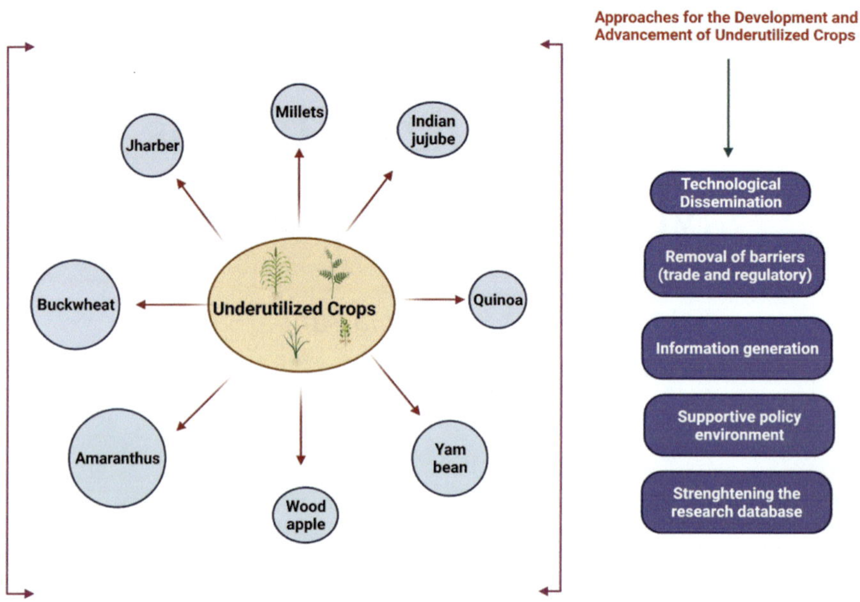

FIGURE 3.1 Underutilized crops with approaches and advancements for their development.

metabolites, they may be used to produce novel antibiotics or disease resistance prototypes [9]. In terms of Jammu and Kashmir, the union territory (UT) is blessed with a vast array of wild edible plants with enormous health-promoting potential [10]. Though the plants have been claimed to have anticancer, antioxidant, anti-inflammatory, and immunity-enhancing effects, that still need to be investigated in the UT of J&K [11].

3.2 UNDERUTILIZED CROPS WITH POTENTIAL NUTRITIONAL AND MEDICINAL VALUES

Human survival, like that of all living organisms, depends on three essential components: glucose, protein, and fat. The three main cereal crops, rice, wheat, and maize, are utilized to meet appropriate nutritional needs, and the bulk of the population heavily depends on them [12]. As a result of the ever-increasing global population, researchers have adopted several transgenic and traditional breeding approaches to change the present farming system. However, concentrating only on boosting global productivity via technology interventions to fulfill food demand, food unpredictability, and dietary requirements remains a problem [13]. With these manipulations, cereal crops become more sensitive to a variety of biotic and abiotic stressors. Thus, underused crops play an important role in addressing a huge portion of the people's nutritional and cultural features [14] (Table 3.1).

TABLE 3.1
Nutritional and Nutraceutical Utility of Different Medicinal Plants

Reference	Major Findings/Outcomes
Asif et al. [9]	The underutilized crops in the immediate future will be exerted into the health and medicine, thus making the food industry research oriented to the pharmaceutical industry.
Ghaffari and Roshanravan [36]	The study demonstrated that nutraceuticals can be utilized for not only reducing cardiovascular risks and maintaining the blood pressure but also can modulate risk factors associated with oxidative stress, hyperlipidemia, and vascular resistance.
Nandini and Sridhara [18]	The study revealed that millets are extremely rich sources of fibers, carbohydrates, and other bioactive compounds.
Jimoh et al. [39]	*Amaranthus caudatus* can be a novel source for the potential therapeutics due to its higher phenolic constituents, flavonoids, and proteins
Maharajan et al. [15]	In developing countries, finger millet can play an essential role in providing the food and nutritional security.
Pathan and Siddiqui [29]	The study found that the Quinoa greens can be utilized for treating wide types of diseases. It can be a potential value-added vegetable that could be an answer to the growing food and nutritional insecurity.
Dabija et al. [34]	Buckwheat and Amaranth pose excellent functional and nutraceutical properties and hence can also be used for high-quality beer production.
Zou et al. [38]	In vitro and in vivo investigation has revealed that buckwheat can offer a plethora of health-enhancing benefits as it contains a wealth of bioactive compounds such as rutin and quercetin.
Essa et al. [35]	The role of functional foods in treating a wide type of cancers and other diseases has been investigated in both in vitro and in vivo as well as same evidences have been noticed using certain clinical studies.
Muthamilarasan et al. [50]	The integration of different omics techniques can provide a notable insight to decipher the different genetic, biological, and other complex traits.

3.2.1 Millets

Millets were the first cereal grain to be used as staple food in various regions, and they are the oldest cereal that humans are familiar with. Millets are particularly nutritious when compared to other main typical cereals. Millet production was pushed back by the introduction of new crop varieties, large external inputs, and increasing food demand during the green revolution [18].

3.2.1.1 Finger Millets (*Eleusine coracana* L.)

Finger millet is a nutrient-dense grain that contains key components, including glucose. It has ten times the calcium content than polished rice [15]. Carbohydrates account for around 80% of the grain of finger millet [16]. It has less fat and protein than most other cereals. The calories in 100 g of finger millet are about 336. It also contains amino acids (valine, isoleucine threonine, methionine, and tryptophan) [17]. Millets are a popular food in developing countries due to its resilience and adaptability to a varied array of climatic conditions [18].

3.2.1.2 Foxtail Millet (*Setaria italic* L.)

Foxtail millet (*Setaria italica*), a round-shaped grain related to maize and sorghum and belonging to the Poaceae family, is available in a range of colors and sizes. It may thrive in any soil type, including tropical and subtropical climates such as Africa and Asia [19]. Biologically active substances found in foxtail millet include dietary fibers, bioactive peptides, proteins, minerals, amino acids, phenolic compounds, and a variety of other nutrients [20]. Foxtail millet is drought resistant and may escape certain droughts due to its early maturity. Because of its quick growth, it may be used as a short-term catch crop [21]. It can thrive in a wide range of heights, soil types, and temperatures. Its grain is used in human diet, poultry feed, and cage bird feed. The phenolic constituents in foxtail millet have been found to be an effective anti-proliferative in case of cancers, particularly in human breast cancer and HepG2 human liver cancer cells [22].

3.2.1.3 Proso Millet (*Panicum millaceum* L.)

Proso millet (*Panicum millaceum*) is also known as common millet, hog millet, and white millet. Grains are nutritionally abundant in essential amino acids and other proteins [23]. In India, proso millet is cooked and consumed like rice. Proso millet is high in antioxidants and vitamin E [24]. It has a protein content that is comparable to wheat and somewhat greater than rice [25]. Furthermore, proso millet-based goods had a lower glycemic response than staple cereal-based products. It has been found that the phenolics in proso millet had anti-diabetic, antioxidant, anti-inflammatory, and antibacterial properties [26].

3.2.2 Quinoa (*Chenopodium quinoa* Willd.)

Because of its particular nutritional value, it is an Andean native that has garnered international attention. Quinoa grains have greater protein amounts and a better distribution of essential amino acids than other cereal grains [27]. It may be used to replace milk protein [28]. The quinoa plant is also resistant to cold, salt, and drought, which is why it is known as the "golden grain." Quinoa grains and quinoa green contain higher concentrations of nutrients to boost health-promoting qualities [29]. Quinoa greens are a high-value vegetable that has the ability to eliminate malnutrition and increase food and nutritional security [30].

3.2.3 Buckwheat (*Fagopyrum*)

Buckwheat is among the most often mentioned pseudocereals in the literature for the creation of gluten-free meals [31]. Buckwheat is a key source of antioxidant chemicals, and its addition to goods has significantly boosted their antioxidant activity.

Consumption of buckwheat on a regular basis might help avoid several "food-related civilization illnesses" (indigestion, obesity, constipation, cholesterol, diabetes, hypertension, and so on) [32]. Buckwheat is gaining popularity as a source of raw materials for functional foods and medicines [33]. Buckwheat is regarded as the "golden culture" of the future by experts owing to its extensive nutraceutical characteristics [34]. Buckwheat contains rutin and poses strong antioxidants, anti-inflammatory, and anti-hypertensive properties, all of which are beneficial in cardiovascular diseases [35]; therefore, its demand is growing in the pharmaceutical industry. Buckwheat flavonoids prevent edema and coronary artery diseases [35], hemorrhagic diseases, and have anti-carcinogenic characteristics [36]. Many buckwheat-based culinary and medicinal items have also been patented across the globe [37]. The Chinese National Intellectual Property Administration, for example, has patented more than 210 traditional dishes, healthful beverages, and medicinal formulations based on *Fagopyrum tataricum* [38].

3.2.4 Amaranth (*Amaranthus*)

Amaranth is a member of the *Amaranthaceae* family that grows in temperate and tropical climates across the globe. Regular consumption of amaranth has shown to have many medicinal properties and can be used to treat several ailments [39]. The potential of the amaranth to deliver health advantages is related to its bioactive substances; for example, rutin has been shown to reduce the aging process, quercetin to prevent oxidation, and nictoflorin to maintain memory functions. Amaranth seed proteins are nutritionally useful, with a high digestibility of around 90% and a percentage of essential amino acids that is near to the World Health Organization (WHO) recommendation [40]. Squalene levels in amaranth oil have been claimed to be the highest in the plant world (at 7%–11% in refined oil). Squalene is strong in antioxidants [41], which protects the skin from oxidative damage caused by free radicals and helps to rejuvenate the skin's protective layer [42]. Amaranth oil is also high in tocotrienols, which help decrease LDL cholesterol and phytosterols that have hypocholesterolemic properties [34].

3.2.5 Yam bean (*Pachyrhizus erosus*)

P. erosus, sometimes known as jicama or yam bean, is an underutilized crop. Production is predicted to expand in the coming years due to its extensive nutrient amounts, nitrogen-fixing characteristics for sustainable farming, and numerous medicinal qualities, as well as its successful introduction in many nations throughout the globe. The presence of constituents such as folates, riboflavin, pyridoxine, pantothenic acid, thiamin, and other phytochemicals in the tubers of *P. erosus* has given birth to its therapeutic potential [43]. The abundance of phytochemicals in seeds and leaves has led to the development of natural medicines and cosmetics [44]. *P. erosus* has been demonstrated in several studies to be effective in treating a number of illnesses and boosting immunity. Some of the identified compounds were revealed to be responsible for *P. erosus* biological activity as an anticancer, antifungal, and antiviral plant [45]. Because they have positive effects against a multitude of illnesses, these properties might be considered the plant's major biological features. Immune modulation has been demonstrated in vivo, in vitro, and ex vivo, but the chemicals responsible for immune modulation are little understood and investigated.

3.2.6 Indian Jujube (*Ziziphus mauritiana* L.)

It thrives in poor or marginal soils, where the majority of commercial fruit trees perish or perform poorly. Saponins, jujubogenin, and obelin lactone are all found in jujube seeds [46]. The fruit is abundant in vitamins C, A, B, carotenoids, protein, Ca, P, K, etc. Its burned leaves produce smoke that is used to cure skin, cough, and cold disorders. As a result, efforts must be made to diversify and promote jujube goods. Due to its outstanding resistance to a wide variety of environmental factors, it is the only fruit crop that can generate considerable yields even under rainfed circumstances.

3.2.7 Wood Apple (*Feronia limonia* L.)

Wood apple is an indigenous fruit tree that thrives as a stray plant capable of surviving both drought and salt stress [47]. The leaves of the *F. limonia* tree have excellent medicinal and functional properties [48].

3.2.8 Jharber (*Ziziphus nummularia Burm.* f.)

Jharber's medicinal properties are used to treat a variety of ailments [48]. Jharber leaves, sometimes called 'Pala,' are rich in crude fiber, crude protein, calcium, and phosphorus that can be utilized as animal fodder. Several plant components are also employed for local medicinal reasons [48]. Different parts of the plant such as leaves, roots, bark, and seeds and their powers are employed for different diseases such as asthma, liver problems, diarrhea, and ulcers [49].

3.3 APPROACHES FOR THE DEVELOPMENT AND ADVANCEMENT OF UNDERUTILIZED CROPS

In rural and tribal areas of the country, where agricultural supply is restricted or non-existent, underutilized crop species can help in alleviating food insecurity. In recent years, certain public research organizations have made considerable efforts to improve underutilized crops. As a result of selection and hybridization, several varieties have been developed for enhanced adaptableness with outstanding harvest and quality traits in dry and semi-arid situations. Stress-related genes discovered in regionally adapted species might be used in future breeding operations to create stress-tolerant variations [48]. End-user training and demonstrations of developing technologies might help promote these underutilized crops more quickly. The focus should be on development activities such as planting nurseries to ensure the provision of high-quality planting materials and processing units to expand their value-added products at suitable levels through entrepreneurship, self-help groups, or farmer producer organizations (FPO), among others [50]. Under the newly announced government supported initiative 'Prakritik kheti' (Natural farming), which aims to promote the production of crops and fruit species in their natural habitats, underused crops are likely to get particular attention. Furthermore, the government has placed a strong focus on the value of indigenous species by using the motto "voice for local."

3.4 CONCLUSION

Due to their great medicinal and nutritional potential, underutilized crops might be regarded future horticultural assets to assist countries in ensuring nutrition and food security, as well as giving recreational, social, and environmental value. These neglected crops might be the next generation of future crops, capable of increasing farmer revenue via sustainable production techniques, especially in the face of climate change. Their horticulture growth is accelerating, and in the future, both traditional horticultural companies and commercial sectors will see more technology adoption, extension, and policy planning. Research and development of appropriate particular packaging procedures and improved kinds will need extra effort. Given the potential of underused crops, more attention should be placed on substantiating efforts to eradicate global hunger by their direct introduction in various parts of the globe, in order to accomplish the United Nations' Sustainable Development Goals (1–3).

Author contribution: Sabreena and Shahnawaz Hassan led the design of this manuscript. The authors read and approve the final manuscript.

Conflict of interest: Authors declare that there is no conflict of interest.

Acknowledgment: The authors are highly thankful to the Department of Environmental Science, University of Kashmir and Division of Plant Biotechnology, SKUAST

REFERENCES

1. Pana, T.A., Bhattacharya, S., Gamble, D.T., Pasdar, Z., Szlachetka, W.A., Perdomo-Lampignano, J.A. (2021). Country-level determinants of the severity of the first global wave of the COVID-19 pandemic: An ecological study. *BMJ Open* 11(2):e042034.
2. Von Grebmer, K., Bernstein, J., Alders, R., Dar, O., Kock, R., Rampa, F. (2020). *Global Hunger Index: One decade to zero hunger: Linking health and sustainable food systems*. Welthungerhilfe: Bonn, Germany.
3. Berwal, M.K., Haldhar, S.M., Ram, C., Shil, S., Kumar, R., Gora, J.S. (2021). Calligonum polygonoides L. as novel source of bioactive compounds in hot arid regions: Evaluation of phytochemical composition and antioxidant activity. *Plants* 10(6):1156.
4. Andreotti, F., Bazile, D., Biaggi, C., Callo-Concha, D., Jacquet, J., Jemal, O.M. (2022). When neglected species gain global interest: Lessons learned from quinoa's boom and bust for teff and minor millet. *Global Food Security* 32:100613.
5. Rodríguez, J.P., Rahman, H., Thushar, S., Singh, R.K. (2020). Healthy and resilient cereals and pseudo-cereals for marginal agriculture: Molecular advances for improving nutrient bioavailability. *Frontiers in Genetics* 11:49.

6. Pandey, A., Malav, P.K., Rai, M.K., Ahlawat, S. (2021). 'Neodomesticates' of the Himalayan allium spices (Allium species) in Uttarakhand, India and studies on eco-geography and morphology. *Genetic Resources and Crop Evolution* 68(5):2167–79.
7. Hiremath, G., Desai, S., Lavanya, V., Patel, N.B, Satisha, T., Biradar, S. (2017). Genetic variability analysis in germplasm collections of buckwheat. *International Journal of Current Microbiology and Applied Sciences* 6:604–710.
8. Murtaza, I., Rashid, N., Laila, O., Ganie, N.A., Wani, M.Y. (2020). Anti-cancer and anti-microbial potential of nutraceutical rich under utilized herb Portulaca Oleracea from Kashmir Region. *International Journal of Current Microbiology and Applied Sciences* 9(9):3496–508.
9. Asif, M., Mohd, I. (2019). Prospects of medicinal plants derived nutraceuticals: A re-emerging new era of medicine and health aid. *Progress in Chemical and Biochemical Research* 2(4):150–69.
10. Nabi, A., Mukarram, M., Aftab, T., Khan, M.M.A., Naeem, M. (2022). Acquisition of physiological modulations in medicinal plants through degraded natural polysaccharides under dynamic environment. In: Tariq Aftab, M. Naeem, (Eds.) *Emerging Plant Growth Regulators in Agriculture*: Elsevier. pp. 399–414. Academic Press.
11. Wali, A., Majid, S., Rasool, S., Shehada, S., Abdulkareem, S., Firdous, A. (2019). Natural products against cancer: Review on phytochemicals from marine sources in preventing cancer. *Saudi Pharmaceutical Journal* 27:767–77.
12. Henchion, M., Hayes, M., Mullen, A.M., Fenelon, M., Tiwari, B. (2017). Future protein supply and demand: Strategies and factors influencing a sustainable equilibrium. *Foods* 6(7):53.
13. King, T., Cole, M., Farber, J.M., Eisenbrand, G., Zabaras, D., Fox, E.M. (2017). Food safety for food security: Relationship between global megatrends and developments in food safety. *Trends in Food Science & Technology* 68:160–75.
14. Mayes, S., Massawe, F., Alderson, P., Roberts, J., Azam-Ali, S., Hermann, M. (2012). The potential for underutilized crops to improve security of food production. *Journal of Experimental Botany* 63(3):1075–9.
15. Maharajan, T., AntonyCeasar, S., Ajeesh Krishna, T.P., Ignacimuthu, S. (2021). Finger millet [Eleusine coracana (L.) Gaertn]: An orphan crop with a potential to alleviate the calcium deficiency in the semi-arid tropics of Asia and Africa. *Frontiers in Sustainable Food Systems* 5:684447. doi: 103389/fsufs.
16. Singh, P., Raghuvanshi, R.S. (2012). Finger millet for food and nutritional security. *African Journal of Food Science* 6(4):77–84.
17. Kabir, M., Oppert, J.M., Vidal, H., Bruzzo, F., Fiquet, C., Wursch, P. (2002). Four-week low-glycemic index breakfast with a modest amount of soluble fibers in type 2 diabetic men. *Metabolism-Clinical and Experimental* 51(7):819–26.
18. Nandini, K., Sridhara, S. (2019). Response of growth yield and quality parameters of foxtail millet genotypes to different planting density. *International Journal of Current Microbiology and Applied Sciences* 8(2):1765–73.
19. Hutabarat, D.J.C., Bowie, V.A. (2022). Bioactive compounds in foxtail millet (Setaria italica)-extraction, biochemical activity, and health functional: A review. *IOP Conference Series:Earth and Environmental Science*. IOP Publishing. 998012060.
20. Dapčević-Hadnađev, T., Hadnađev, M., Pojić, M. (2018). The healthy components of cereal by-products and their functional properties. In: Charis M. Galanakis (Ed.) *Sustainable Recovery and Reutilization of Cereal Processing by-Products*. Elsevier. pp. 27–61.
21. Chivenge, P., Mabhaudhi, T., Modi, A.T., Mafongoya, P. (2015). The potential role of neglected and underutilised crop species as future crops under water scarce conditions in Sub-Saharan Africa. *International Journal of Environmental Research and Public Health* 12(6):5685–711.

22. Wassmann, R., Jagadish, S., Heuer, S., Ismail, A., Redona, E., Serraj, R. (2009). Climate change affecting rice production: The physiological and agronomic basis for possible adaptation strategies. *Advances in Agronomy* 101:59–122.
23. Gomashe, S.S. (2017). Proso Millet, Panicum miliaceum (L.): Genetic improvement and research needs. In: J. V. Patil (Ed.) *Millets and Sorghum: Biology and Genetic Improvement*. Wiley. pp. 150–69.
24. Das, S., Khound, R., Santra, M., Santra, D. K. (2019). Beyond bird feed: Proso millet for human health and environment. *Agriculture* 9(3):64.
25. Verma, S., Srivastava, S., Tiwari, N. (2015). Comparative study on nutritional and sensory quality of barnyard and foxtail millet food products with traditional rice products. *Journal of Food Science and Technology* 52(8):5147–55.
26. Chandra, D., Chandra, S., Sharma, A. (2016). Review of Finger millet (Eleusine coracana (L.) Gaertn): A power house of health benefiting nutrients. *Food Science and Human Wellness* 5(3):149–55.
27. Angeli, V., Miguel Silva, P., Crispim Massuela, D., Khan, M.W., Hamar, A., Khajehei, F. (2020). Quinoa (Chenopodium quinoa Willd.): An overview of the potentials of the "golden grain" and socio-economic and environmental aspects of its cultivation and marketization. *Foods* 9(2):216.
28. Filho, A.M.M., Pirozi, M.R., Borges, J.T.D.S., Pinheiro Sant'Ana, H.M., Chaves, J.B.P, Coimbra, J.S.D.R. (2017). Quinoa: Nutritional, functional, and antinutritional aspects. *Critical Reviews in Food Science and Nutrition* 57(8):1618–30.
29. Pathan, S., Siddiqui, R.A. (2022). Nutritional composition and bioactive components in Quinoa (Chenopodium quinoa Willd.) Greens: A review. *Nutrients* 14(3):558.
30. Fasolin, L.H., Pereira, R.N., Pinheiro, A.C., Martins, J.T., Andrade, C., Ramos, O. (2019). Emergent food proteins–Towards sustainability, health and innovation. *Food Research International* 125:108586.
31. Zini, P.B., Poletto, T., Fantinel, V.S., Andrade, N.d., Nunes, U.R., Muniz, Md.F.B. (2022). Buckwheat seed quality and pathogenicity of Fusarium spp. in plants. *Journal of Seed Science* 44, e202244004.
32. Cadenas, R., Caballero, I., Nimubona, D., Blanco, C.A. (2021). Brewing with starchy adjuncts: Its influence on the sensory and nutritional properties of beer. *Foods* 10(8):1726.
33. Chusova, A., Romanyuk, T., Agafonov, G., Tararykov, M., Zueva, N. (2021). Selection of optimal conditions for obtaining malt from buckwheat. *Proceedings of the Voronezh State University of Engineering Technologies* 83(2):93–101.
34. Dabija, A., Ciocan, M.E., Chetrariu, A., Codină G.G. (2022). Buckwheat and Amaranth as raw materials for brewing, a review. *Plants* 11(6):756.
35. Essa, M.M., Bishir, M., Bhat, A., Chidambaram, S.B., Al-Balushi, B., Hamdan, H. (2021). Functional foods and their impact on health. *Journal of Food Science and Technology* 60:820–834.
36. Ghaffari, S., Roshanravan, N. (2020). The role of nutraceuticals in prevention and treatment of hypertension: An updated review of the literature. *Food Research International* 128:108749.
37. Giménez-Bastida, J.A., Piskula, M.K., Zielinski, H. (2015). Recent advances in processing and development of buckwheat derived bakery and non-bakery products-a review. *Polish Journal of Food and Nutrition Sciences* 65(1):9–20.
38. Zou, L., Wu, D., Ren, G., Hu, Y., Peng, L., Zhao, J. (2021). Bioactive compounds, health benefits, and industrial applications of Tartary buckwheat (Fagopyrum tataricum). *Critical Reviews in Food Science and Nutrition* 1:31–47.
39. Jimoh, M., Afolayan, A., Lewu, F. (2019). Therapeutic uses of Amaranthus caudatus L. *Tropical Biomedicine* 36:1038–53.

40. Malik, A.M., Singh, A., Nanda, V. (2021). Pseudocereals proteins-A comprehensive review on its isolation, composition and quality evaluation techniques. *Food Chemistry Advances* 1:100001.
41. Karaca, A.C. (2022). Proteins from pseudocereal grains. *Pseudocereals*, IntechOpen. p. 35–44.
42. Ibrahim, N.I., Fairus, S., Zulfarina, M.S., Naina, Mohamed, I. (2020). The efficacy of squalene in cardiovascular disease risk – A systematic review. *Nutrients* 12(2):414.
43. Sujono, T.A., Nurrochmad, A., Lukitaningsih, E., Nugroho, A.E. (2021). Immunomodulatory effect of methanolic extract and ethyl acetate fraction of Bengkoang (Pachyrhizus erosus (L.) urban) tuber in mice. *Pharmacognosy Journal* 13(1):59–65.
44. Lee, A., Kim, G.N., Kim, H.O., Song, W., Roh, S.S. (2017). Antioxidant activity and melanin inhibitory effects of yambean (Pachyrhizus erosus) extract. *The Korea Journal of Herbology* 32(2):57–64.
45. Jaiswal, V., Chauhan, S., Lee, H.J. (2021). The bioactivity and phytochemicals of Pachyrhizus erosus (L.) Urb.: A multifunctional underutilized crop plant. *Antioxidants* 11(1):58.
46. Delfanian, M., Esmaeilzadeh Kenari, R., Sahari, M.A. (2016). Utilization of Jujube fruit (Ziziphus mauritiana Lam.) extracts as natural antioxidants in stability of frying oil. *International Journal of Food Properties* 19(4):789–801.
47. Elias, E.H., Flynn, R., Idowu, O.J., Reyes, J., Sanogo, S., Schutte, B.J. (2019). Crop vulnerability to weather and climate risk: Analysis of interacting systems and adaptation efficacy for sustainable crop production. *Sustainability* 11(23):6619.
48. Meena, V.S., Gora, J.S., Singh, A., Ram, C., Meena, N.K., Rouphael, Y. (2022). Underutilized fruit crops of Indian arid and semi-arid regions: Importance, conservation and utilization strategies. *Horticulturae* 8(2):171.
49. Sivalingam, P., Mahajan, M.M., Satheesh, V., Chauhan, S., Changal, H., Gurjar, K. (2021). Distinct morpho-physiological and biochemical features of arid and hyper-arid ecotypes of Ziziphus nummularia under drought suggest its higher tolerance compared with semi-arid ecotype. *Tree Physiology* 41(11):2063–81.
50. Muthamilarasan, M., Singh, N.K., Prasad, M. (2019). Multi-omics approaches for strategic improvement of stress tolerance in underutilized crop species: A climate change perspective. *Advances in Genetics* 103:1–38.

4 Buckwheat and Cancer

Sabreena, Shahnawaz Hassan, and Bashir Ah Ganai
University of Kashmir, Srinagar, India

4.1 INTRODUCTION

Cancer is recognized as one of the most lethal diseases worldwide. Cancer mortality is on the upsurge trend and has become one of the foremost bases of human mortality [1]. Although significant advancement has been made in the control and treatment of cancer, it still caused approximately 9.9 million deaths in 2020. Different conventional treatments such as chemotherapy or its amalgamation with radiotherapy have been put in place to treat such disorders. However, there are certain apprehensions with the usage of such techniques as they can lead to other critical disorders. Consequently, in many cancer models, naturally occurring anti-cancer agents have been studied in both in vitro and in vivo [2]. Similarly, various plant-based products employed in the clinical manifestations have shown good results without promoting any antagonistic effects [3]. There are unexplored genetic resources that have extremely high potential to address the deteriorating health scenario across the world. The crop germplasm with diverse data and population structure can be a valuable genetic source to discover the potential genes to treat deadly diseases such as cancer. Buckwheat (*Fagopyrum*) is one such traditional medicinal plant of the *Polygonaceae* family [4]. Important species such as common buckwheat (*Fagopyrum esculentum* Moench), Tartary buckwheat (TB) (*Fagopyrum tataricum* (L.) Gaertn.), and golden or tall buckwheat (*Fagopyrum dibotrys*) [5] have gained a tremendous significance in agricultural arena as a possible "functional food" source all over the world [6]. These buckwheat species have different morphological characteristics. Scientific investigations have reported that buckwheat species contain abundant proteins and phenolics, making these potential therapeutic plants [7]. Out of all buckwheat species listed above, *F. tataricum* has been stated to encompass a higher concentration of phenolics as compared to *F. esculentum* [8]. Due to these properties, it has been under serious scientific investigation [9]. Besides being edible crops, buckwheat species also have been used in medicine, particularly in China, for their ability to invigorate the spleen, reducing food stagnation. Similarly, in British Herbal Pharmacopoeia and Korean folk medicine, buckwheat is being utilized as an anti-hemorrhagic, hypotensive, anti-inflammatory, antioxidant, anti-tumor, anti-cancer, anti-diabetic, anti-obesity, hypoglycemic, and hypo-lipidemic purposes [10,11]. *F. tataricum* roots were extensively used to cure a variety of chronic and incurable ailments, including rheumatic disorders and malignancies. The significant aspect of utilizing plant-based extracts against cancer is that they provide a synergistic behavior and polypharmacological methods for cancer therapies owing to their multi-targeted means of action. Plants

DOI: 10.1201/9781003477792-4

such as buckwheat contain a rich amount of secondary metabolites that have influence on cell cycle and potential anti-cancer properties [12]. This has prompted the researchers to explore the bioactive metabolites from the roots of *F. tataricum*, which led to the discovery of a large number of cytotoxic phenylpropanoid glycosides with a potential for cancer therapy [7]. Flavonoids, phenolics, fagopyritols, triterpenoids, steroids, and fatty acids are the six primary types of buckwheat metabolites [13]. The main active components were thought to be flavonoids and phenolic chemicals [14]. As a result, bioactive molecules like these have been identified as the primary cause of TB's health-promoting benefits [13,15].

4.2 BUCKWHEAT AS SOURCE BIOACTIVE METABOLITES AND OTHER PHENOLIC COMPOUNDS (TABLE 4.1)

4.2.1 Chlorogenic Acid

Chlorogenic acid is a hydroxycinnamic acid that occurs naturally. It's a crucial biochemical intermediary [16]. In the production of chlorogenic acid, the polyphenolic acids, caffeic acid and cyclitol (-)-quinic acid, act as esters [17,18]. Chlorogenic acid has recently caught our attention as it has chemo-preventive properties. Chlorogenic acid, which contains microsomal G6PT (glucose-6-phosphate translocase), is a dominant functional inhibitor with anti-cancer potential [19,20]. It's also a prospective precursor chemical for therapeutic research against the HIV virus [21]. Buckwheat seedlings contain higher concentration of chlorogenic acid [22,24]. Investigations

TABLE 4.1
A Sample of Research Publications Depicting the Medicinal Utility of Buckwheat Phytochemicals

Reference	Plant Part Used	Main Findings/Outcomes
Mukhtar et al. [9]	Leaves, seeds, and roots	With the intervention of new biotechnological approaches, it is possible to improve the genetic variability of traits in Buckwheat. This could have serious implications with respect to various disease treatments.
Rodriguez et al. [10]	Leaves	Buckwheat contains an abundance of molecules that can be utilized for the treatment of hepatocellular carcinoma.
Zou et al. [13]	Leaves, seeds, and roots	Clinical manifestations of buckwheat have revealed that due to an abundance of bioactive compounds, Tartary buckwheat unveils significant anti-cancer properties.

(*Continued*)

TABLE 4.1 (*Continued*)
A Sample of Research Publications Depicting the Medicinal Utility of Buckwheat Phytochemicals

Reference	Plant Part Used	Main Findings/Outcomes
Huda et al. [29]	Leaves, seeds, and roots	The presence of flavonoids in buckwheat has significant health benefits and can be utilized as an anti-cancer agent. However, it is important to conduct more studies to validate and isolate the unexplored genes involved in the biosynthetic pathways.
Vollmannová et al. [32]	Leaves, flowers, and grains	Phenolic compound abundance in buckwheat plant has diverse pharmaceutical potentials.
Zhang et al. [56]	Leaves, flowers, seeds, and roots	The study revealed that *Fagopyrum dibotrys* posseses significant anti-cancer properties.
Ruan et al. [4]	Leaves, seeds, and roots	The study demonstrated that the occurrence of flavonoids effectively prevents cell damage by quenching the free radicals.
Li et al. [49]	Tartary buckwheat bran	The study involved the stream explosion and revealed the anti-proliferative property of the phenolic extract against HepG2 cells.
Sytar et al. [28]	Leaves, flowers, seeds, and roots	Buckwheat contains a good concentration of phenolic compounds and can be applied in disease treatments due to its high antioxidant capacity.
Jeong et al. [51]	Buckwheat sprouts	The study revealed that using buckwheat sprouts on HepG2 cells considerably decreased the oxidative damage as it modulated the effects of free radicals.

have reported that different nickel (Ni) concentrations alter chlorogenic acid concentration in buckwheat [25].

4.2.2 Fagopyrin

Fagopyrin is a naphthodianthrone with anti-cancer properties. Fagopyrin was initially isolated from *F. esculentum*'s red flowered variety in 1943. Fagopyrin has a structure comparable to hypericin from St. John's wort (*Hypericum perforatum* L.). The action of hypericin and fagopyrin is light-dependent. Because of this, they are employed in medicine for photodynamic treatment [26]. According to studies, fagopyrin inhibits protein tyrosine kinase (PTK), a protein that regulates cancer cell growth and so has a substantial anti-carcinogenic effect [27,28]. There are currently little data on the buildup of fagopyrin in various parts of buckwheat [29]. It has been reported that flowers in buckwheat contain the highest fagopyrin content (0.08% FW) as compared to leaves (0.05% FW) [30], which in turn has greater concentration (2–3) times as compared to stems [31,32].

4.2.3 Phenylpropanoid Glycosides

Traditional Chinese medicine has used TB root to treat chronic ailments such as rheumatoid arthritis and cancer [33]. Using ethyl acetate, phenylpropanoid glycosides in TB root was isolated and characterized using mass spectrophotometry (MS) and nuclear magnetic resonance (NMR) methods [34,23]. Different analytical techniques have been used such as high-performance liquid chromatography with photodiode array detector (HPLC–PDA) and linear ion trap Fourier transform ion cyclotron resonance mass spectroscopy (LTQ–FTICR–MS) to examine the composition of phenylpropanoid glycosides in TB seeds [35]. In vitro bioactivity of these phenylpropanoid glycosides was investigated for anti-cancer activity. The cytotoxic effects of buckwheat extract's phenylpropanoid glycosides have been studied on many human cancer cell lines. Tatariside C (3), with an IC50 of 6.44–7.49 lg/mL against cancer cell lines, has been found to be the most active among phenylpropanoid glycosides [34].

4.2.4 Polyphenols

It is widely known that high polyphenol consumption in the diet is linked to a lower risk of chronic illnesses, including cancer, neurodegeneration, and cardiovascular disease. In TB, polyphenols are the most investigated phytochemicals. According to recent research, long-term consumption of polyphenol-rich diets may protect against cancer, cardiovascular disease, and neurological issues [36,37]. This can be attributed to biochemical scavenging, in which free radicals are neutralized by the polyphenols making stable complexes to stop any further reactions [38]. Polyphenols also defend against oxidative stress by producing hydrogen peroxide (H_2O_2), which may affect immune response functions including cellular growth [38]. However, much of the study is done in vitro, and it is uncertain if these processes can be transferred to societal benefit [39]. Furthermore, recent evidence has elucidated the effect of absorption pharmacokinetics on efficacy of polyphenols as antioxidants and other potentially health-promoting mechanisms; these physiochemical properties of the molecules may explain the variable effects observed in human and animal models, as well as conflicting data in the literature [40,41].

4.2.5 Proteins

Buckwheat proteins (BWPs) are known for their anti-cancer properties. Buckwheat protease inhibitors (BWI-1 and BWI338 2a) and their recombinant variants (BWI-1 and BWI338 2a) have exhibited anti-cancer action in a range of cancer cell types by causing apoptosis and inhibiting proliferation [42]. New BWPs (TBWSP31) with anti-cancer action have been discovered in recent investigations. Similarly, proliferation of HepG2 cells in a human cancer cell line and liver embryonic cell lines showed inhibition by 341 peptides from buckwheat (BW) seeds [43]. Protein TBWSP31 is an anti-tumor protein isolated from TB and possesses anti-cancer and anti-proliferative actions while applied in different dose ranges (5–200 g/mL) and prompts apoptosis in human breast cancer cell Bcap37 [44].

More in vitro and in vivo investigations are needed to confirm the possible medicinal activities of BW and its constituent components, despite the fact that these findings open the door to novel medicines for disease prevention and treatment.

4.3 BUCKWHEAT AND CANCER

Cancer is the most common cause of mortality, affecting both rich and poor. One of the century's major issues is the development of functional foods for the prevention of chronic illnesses like cancer. Consuming a variety of meals, including BW (buckwheat), on a regular basis lowers the incidence of lung cancer [45]. Because oxidative DNA damage is linked to cancer genesis and progression, BW's anti-cancer activities have been linked to its antioxidant qualities. In the HepG2 cell line, buckwheat extracts have demonstrated a potent antioxidant capacity 2,2-diphenyl-1-picrylhydrazyl (DPPH scavenging activity) that prevents tert-butyl hydroperoxide from causing DNA damage [46]. Changes in gene expression, decrease of proliferation, cell cycle arrest, and/or activation of apoptosis have all been found as means of action in vivo and in vitro, in addition to antioxidant capabilities as shown in Figures 4.1 and 4.2.

4.3.1 Anti-Cancer Activity of *F. tataricum*

Anti-cancer efficacy of buckwheat extracts and its components has been studied extensively in vitro and in vivo [8,47]. Various components of buckwheat, including protein [44], polyphenols [48], and phenylpropanoid glycosides [34], have revealed the cytotoxicity on assorted cancerous cell lines [49]. Similarly, the molecular and genetic effects of protein (TBWSP31) and polyphenols on malignant cells have been investigated. The study demonstrated that TBWSP3, on the human breast cancer cell line (Bcap37), brought morphological changes and further analysis by scanning electron microscopy revealed the presence of potential anti-tumor compounds [44,48]. Similarly, the occurrence of different phenolics in TB was found to exhibit anti-cancer properties against breast and gastric cancer [50] mediated by enhanced production of caspase-8 protein [13]. Quercetin obtained from TB has been reported to exhibit cytotoxic effects on human hepatomaHepG2 cells. Furthermore, TB-derived quercetin has been shown to have potent cytotoxic effects and trigger cell cycle arrest at the G2/M phase, ensuing a surge in apoptotic cell death [48]. Indeed, quercetin can reduce human breast cancer's motility [51]. Furthermore, through enhancing caspase-3 and caspase-9 activity, quercetin improves the efficiency of gemcitabine-induced cell apoptosis in lung cancer A549 and H460 cells [52]. In THP-1 human leukemia cells, TB polysaccharides, in addition to polyphenols, exhibit anti-cancer properties, stimulating cell differentiation and maturation and

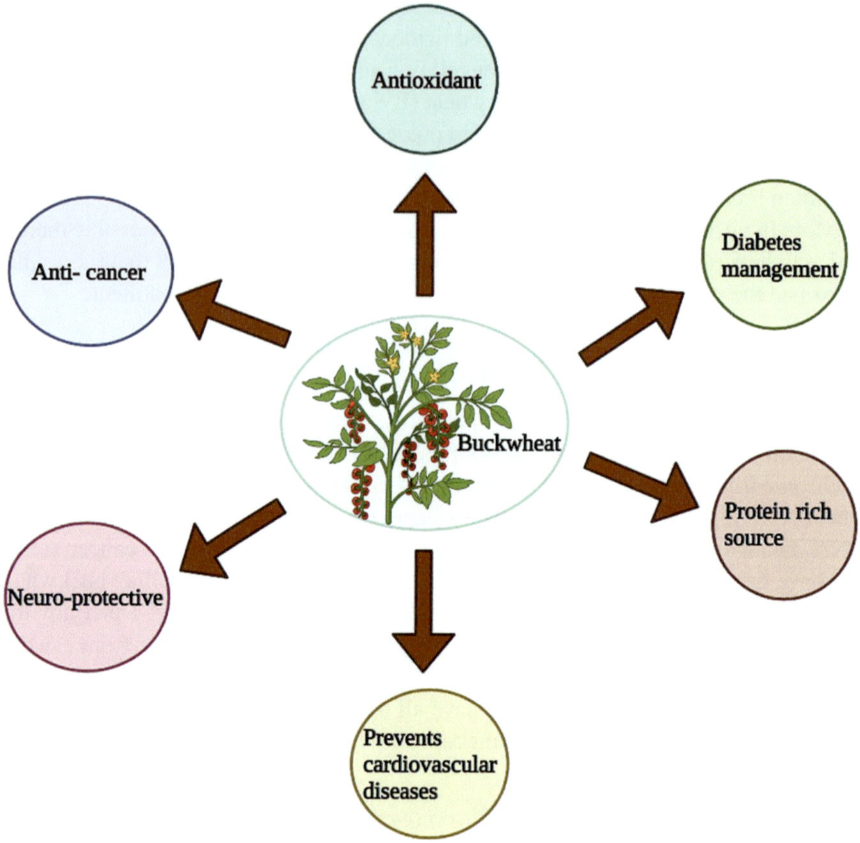

FIGURE 4.1 Buckwheat and its medicinal utilities.

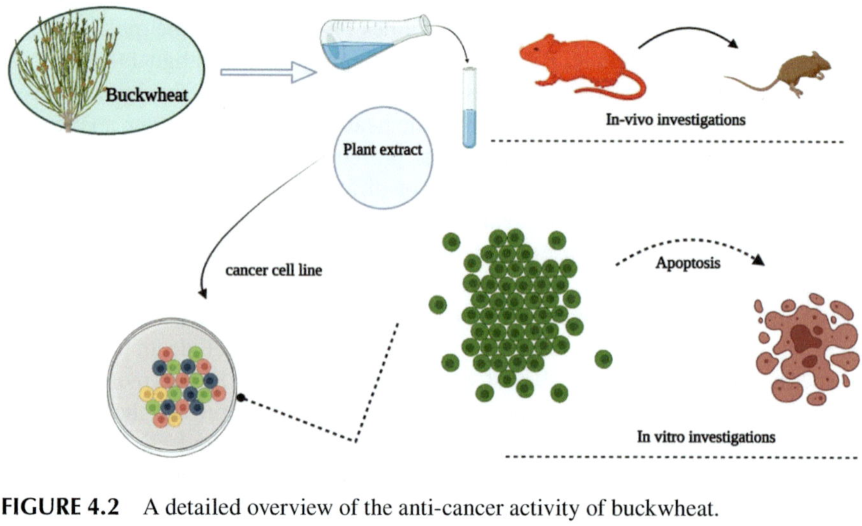

FIGURE 4.2 A detailed overview of the anti-cancer activity of buckwheat.

increasing phagocytic activity in mature cells [53]. In addition, BWP (TB protein) attenuated the effects of 7,12-dimethylbenz[α]anthracene-induced breast carcinogenesis in rats [54] by suppressing cell proliferation and having anti-cancer activity [55]. These investigations provide a plethora of information on the anti-cancer activities of TB polyphenols, polysaccharides, and proteins in a broad range of cancer cell lines.

4.3.2 Anti-Cancer Activity of *F. dibotrys*

F. dibotrys (*Fagopyrum cymosum*) has long been utilized in China to cure a range of lung diseases. In vitro tests found that commercial preparations of *F. cymosum* exhibited broad-spectrum cytotoxicity, significantly suppressing cancer cell proliferation. It not only has anti-cancer properties, but its extracts also have anti-proliferative properties [56]. Fr4 is the most promising fraction (the fourth fraction from the extract of *F.* cymosum rhizomes). This fraction of buckwheat, administered at a dosage of 400 mg/kg to mice, reduces lung cancers and lowers matrix metalloprotease (MMP-9) synthesis [56]. In vitro combination of *F. cymosum* and *Rosa roxburghii* extracts had a synergistic effect in suppressing proliferation and inducing cell death [57]. Similarly, the combination of *F. cymosum* extracts and matrine inhibited expression, resulting in a reduction in lung cancer cell line metastasis (HepG2) [58].

4.3.3 Anti-Cancer Activity of *Fagopyrum esculentum*

An extract of *F. esculentum* flowers and leaves (EBFL) inhibits tumor cell growth and increases apoptosis in H22 cells (murine hepatic carcinoma cells). Furthermore, by enhancing the immunological function of H22 tumor mice, it might ameliorate cyclophosphamide (CTX)-induced immune suppression [8]. The extract from the plant stopped HL-60 cells from progressing from G0/G1 to S phase in vitro [8]. The ethyl acetate and butanol fractions of *F. esculentum* sprout ethanol extract showed high cytotoxicity against cancer cell lines at a dosage of 1.0 mg/mL [59]. Furthermore, rBTI (recombinant buckwheat trypsin inhibitor) showed substantially greater anti-proliferative effectiveness in vitro against HL-60, EC9706, and HepG2 cells, suggesting that it might be a novel cancer therapeutic option [8]. Buckwheat consumption has been linked to a lower incidence of lung cancer in a population-based case-control study [60].

4.4 CONCLUSIONS AND FUTURE OUTLOOK

Buckwheat contains a unique composition of bioactive components in all of its forms (hull, seed, leaf, blossom, and sprout). Buckwheat has a high concentration of polyphenols (especially flavonols). Compounds such as N-trans feruloyltyramine, 200-hydroxynicotianamin and nicotinamide, and phenylpropanoid glycosides are examples of phytochemicals that are present in modest amounts yet have medicinal and nutritional value. It has a high potential to positively affect human health as a consequence of its unique chemical makeup, as evidenced by in vitro and in vivo experiments as well as an epidemiological study particularly as anti-cancerous. TB extract appears to have greater bioactivities than the pure chemical substance, suggesting that the chemicals interact synergistically and that these extracts can be used in practical applications. Buckwheat compounds such as fagopyrin and protein allergens may have a deleterious influence on human health. TB contains more polyphenols and has greater bioactivity than regular buckwheat. Furthermore, different markers can be considered for polymorphism and association mapping in buckwheat to characterize economically significant traits [61]. According to the results of scientific studies, buckwheat has anti-cancer capabilities against a variety of different cancer cell types.

Author contribution: Sabreena and Shahnawaz Hassan led the design of this manuscript. The authors read and approved the final manuscript.

Conflict of interest: The authors declared that there is no conflict of interest.

Acknowledgment: The authors are highly thankful to the Department of Environmental Science, University of Kashmir, and Division of Plant Biotechnology, SKUAST.

REFERENCES

1. Choi, J.R., Kozalak, G., di Bari, I., Babar, Q., Niknam, Z., Rasmi, Y., & Yong, K.W. (2022). In vitro human cancer models for biomedical applications. *Cancers, 14*, 2284.
2. Rufino-Palomares, E.E., Pérez-Jiménez, A., García-Salguero, L., Mokhtari, K., Reyes-Zurita, F.J., Peragón-Sánchez, J., & Lupiáñez, J.A. (2022). Nutraceutical role of polyphenols and triterpenes present in the extracts of fruits and leaves of Olea europaea as antioxidants, anti-infectives and anticancer agents on healthy growth. *Molecules, 27*, 2341.
3. Khan, A.W., Farooq, M., Haseeb, M., & Choi, S. (2022). Role of plant-derived active constituents in cancer treatment and their mechanisms of action. *Cells, 11*, 1326.
4. Ruan, J., Zhou, Y., Yan, J., Zhou, M., Woo, S.-H., Weng, W., Cheng, J., & Zhang, K. (2020). Tartary buckwheat: An under-utilized edible and medicinal herb for food and nutritional security. *Food Reviews International, 38*, 440–454.
5. Zhu, F. (2016). Buckwheat starch: Structures, properties, and applications. *Trends in Food Science & Technology, 49*, 121–135.
6. Li, J., Yang, P., Yang, Q., Gong, X., Ma, H., Dang, K., Chen, G., Gao, X., & Feng, B. (2019). Analysis of flavonoid metabolites in buckwheat leaves using UPLC-ESI-MS/MS. *Molecules, 24*(7), 1310.
7. Joshi, D., Chaudhari, G.V., Sood, S., Kant, L., Pattanayak, A., Zhang, K., Fan, Y., Janovská, D., Meglič, V., & Zhou, M. (2019). Revisiting the versatile buckwheat: Reinvigorating genetic gains through integrated breeding and genomics approach. *Planta, 250*(3), 783–801.
8. Jing, R., Li, H.-Q., Hu, C.-L., Jiang, Y.-P., Qin, L.-P., & Zheng, C.-J. (2016). Phytochemical and pharmacological profiles of three Fagopyrum buckwheats. *International Journal of Molecular Sciences, 17*(4), 589.
9. Mukhtar, S., Bashir, Z., Mir, R.A., & Zargar, S.M. (2021). Genomic approaches for the improvement and conservation of buckwheat. In: Sajad Majeed Zargar, Antonio Masi, & Romesh Kumar Salgotra (Eds.) *Neglected and Underutilized Crops-Towards Nutritional Security and Sustainability* (pp. 167–180). Springer.
10. Rodriguez, S., Skeet, K., Mehmetoglu-Gurbuz, T., Goldfarb, M., Karri, S., Rocha, J., Shahinian, M., Yazadi, A., Poudel, S., & Subramani, R. (2021). Phytochemicals as an alternative or integrative option, in conjunction with conventional treatments for hepatocellular carcinoma. Cancers, *13*(22), 5753.

11. Sonkamble, V.V., Wagh, N.S., & Pai, S.R. (2019). Role of plant secondary metabolites as antidiabetic agents. In: Mohd Sayeed Akhtar, Mallappa Kumara Swamy, & Uma Rani Sinniah (Eds.) *Natural Bio-Active Compounds* (pp. 529–550). Springer.
12. Truong, V.L., & Jeong, W.S. (2021). Cellular defensive mechanisms of tea polyphenols: Structure-activity relationship. *International Journal of Molecular Sciences, 22*, 9109.
13. Zou, L., Wu, D., Ren, G., Hu, Y., Peng, L., Zhao, J., Garcia-Perez, P., Carpena, M., Prieto, M.A., & Cao, H. (2021). Bioactive compounds, health benefits, and industrial applications of Tartary buckwheat (Fagopyrum tataricum). *Critical Reviews in Food Science and Nutrition*, 63(5), 657–73.
14. Hussein, R.A., & El-Anssary, A.A. (2019). Plants secondary metabolites: The key drivers of the pharmacological actions of medicinal plants. Herbal Medicine, *1*, 13.
15. Zhou, Y., Wang, T., She, X., Wei, Y., Zhou, X., & Xiao, Y. (2022). Pretreatment of buckwheat globulin by ultra-high pressure: Effects on enzymatic hydrolysis and final hydrolysate lipid metabolism regulation capacities. Food Chemistry, *379*, 132102.
16. Flourat, A.L., Combes, J., Bailly-Maitre-Grand, C., Magnien, K., Haudrechy, A., Renault, J.H., & Allais, F. (2021). Accessing P-hydroxycinnamic acids: Chemical synthesis, biomass recovery, or engineered microbial production? ChemSusChem, *14*(1), 118–129.
17. Clifford, M.N., Jaganath, I.B., Ludwig, I.A., & Crozier, A. (2017). Chlorogenic acids and the acyl-quinic acids: Discovery, biosynthesis, bioavailability and bioactivity. Natural Product Reports, *34*(12), 1391–1421.
18. Naveed, M., Hejazi, V., Abbas, M., Kamboh, A.A., Khan, G.J., Shumzaid, M., Ahmad, F., Babazadeh, D., FangFang, X., & Modarresi-Ghazani, F. (2018). Chlorogenic acid (CGA): A pharmacological review and call for further research. *Biomedicine & Pharmacotherapy, 97*, 67–74.
19. Belkaid, A., Currie, J.-C., Desgagnés, J., & Annabi, B. (2006). The chemopreventive properties of chlorogenic acid reveal a potential new role for the microsomal glucose-6-phosphate translocase in brain tumor progression. Cancer Cell International, *6*(1), 1–12.
20. Yan, Y., Li, J., Han, J., Hou, N., Song, Y., & Dong, L. (2015). Chlorogenic acid enhances the effects of 5-fluorouracil in human hepatocellular carcinoma cells through the inhibition of extracellular signal-regulated kinases. Anti-Cancer Drugs, *26*(5), 540.
21. Khan, F., Bamunuarachchi, N.I., Tabassum, N., & Kim, Y.-M. (2021). Caffeic acid and its derivatives: Antimicrobial drugs toward microbial pathogens. *Journal of Agricultural and Food Chemistry, 69*(10), 2979–3004.
22. Adamczak, A., Ożarowski, M., & Karpiński, T.M. (2019). Antibacterial activity of some flavonoids and organic acids widely distributed in plants. *Journal of Clinical Medicine, 9*(1), 109.
23. Oksana, S., Marian, B., Mahendra, R., & Bo, S.H. (2012). Plant phenolic compounds for food, pharmaceutical and cosmetics production. *Journal of Medicinal Plants Research, 6*(13), 2526–2539.
24. Sytar, O., Brestic, M., Zivcak, M., & Phan Tran, L.-S. (2016). The contribution of buckwheat genetic resources to health and dietary diversity. Current Genomics, *17*(3), 193–206.
25. Borovaya, S., & Klykov, A. (2020). Some aspects of flavonoid biosynthesis and accumulation in buckwheat plants. *Plant Biotechnology Reports, 14*(2), 213–225.
26. Mariewskaya, K.A., Tyurin, A.P., Chistov, A.A., Korshun, V.A., Alferova, V.A., & Ustinov, A.V. (2021). Photosensitizing antivirals. *Molecules, 26*(13), 3971.
27. Samel, D., & de Witte, P. (1994). Isolation and PTK inhibitory activity of fagopyrins from buckwheat. In *Papers and Posters 19th LOF-Symposium on Pharmacognosy and Natural Products Chemistry, Pharmacy World & Science.* 16(9), pp. 13–113.

28. Sytar, O., Biel, W., Smetanska, I., & Brestic, M. (2018). Bioactive compounds and their biofunctional properties of different buckwheat germplasms for food processing. In: Meiliang Zhou, Galina Suvorova, Sun Hee Woo, Ivan Kreft, &Yu Tang (Eds.) *Buckwheat Germplasm in the World* (pp. 191–204). Elsevier.
29. Huda, M.N., Lu, S., Jahan, T., Ding, M., Jha, R., Zhang, K., Zhang, W., Georgiev, M.I., Park, S.U., & Zhou, M. (2021). Treasure from garden: Bioactive compounds of buckwheat. *Food Chemistry, 335*, 127653.
30. Sytar, O., Brestic, M., & M. Rai. (2013). Possible ways of fagopyrin biosynthesis and production in buckwheat plants. *Fitoterapia, 84*, 72–79.
31. Ožbolt, L., Kreft, S., Kreft, I., Germ, M., & Stibilj, V. (2008). Distribution of selenium and phenolics in buckwheat plants grown from seeds soaked in Se solution and under different levels of UV-B radiation. *Food Chemistry, 110*(3), 691–696.
32. Vollmannová, A., Musilová, J., Lidiková, J., Árvay, J., Šnirc, M., Tóth, T., Bojňanská, T., Čičová, I., Kreft, I., & Germ, M. (2021). Concentrations of phenolic acids are differently genetically determined in leaves, flowers, and grain of common buckwheat (Fagopyrum esculentum Moench). Plants, *10*(6), 1142.
33. Zhao, J.-S., Yu, C.-H., & Peng, C. (2017). Fagopyrum Mill-A review of phytochemistry and pharmacological effects. In *2017 2nd International Conference on Materials Science, Machinery and Energy Engineering (MSMEE 2017).* (pp. 72–81). Atlantis Press.
34. Zheng, C., Hu, C., Ma, X., Peng, C., Zhang, H., & Qin, L. (2012). Cytotoxic phenylpropanoid glycosides from Fagopyrum tataricum (L.) Gaertn. *Food Chemistry, 132*(1), 433–438.
35. Ren, Q., Wu, C., Ren, Y., & Zhang, J. (2013). Characterization and identification of the chemical constituents from tartary buckwheat (Fagopyrum tataricum Gaertn) by high performance liquid chromatography/photodiode array detector/linear ion trap FTICR hybrid mass spectrometry. Food Chemistry, *136*(3–4), 1377–1389.
36. Fujiki, H., Sueoka, E., Watanabe, T., & Suganuma, M. (2015). Primary cancer prevention by green tea, and tertiary cancer prevention by the combination of green tea catechins and anticancer compounds. *Journal of Cancer Prevention, 20*(1), 1.
37. Xiao, J., & Hogger, P. (2015). Dietary polyphenols and type 2 diabetes: Current insights and future perspectives. *Current Medicinal Chemistry, 22*(1), 23–38.
38. Saeidnia, S., & Abdollahi, M. (2013). Antioxidants: Friends or foe in prevention or treatment of cancer: The debate of the century. *Toxicology and Applied Pharmacology, 271*(1), 49–63.
39. Habas, K., Brinkworth, M.H., & Anderson, D. (2017). Diethylstilbestrol induces oxidative DNA damage, resulting in apoptosis of spermatogonial stem cells in vitro. *Toxicology, 382*, 117–121.
40. Selby-Pham, S.N., Cottrell, J.J., Dunshea, F.R., Ng, K., Bennett, L.E., & Howell, K.S. (2017). Dietary phytochemicals promote health by enhancing antioxidant defence in a pig model. *Nutrients, 9*(7), 758.
41. Selby-Pham, S.N., Howell, K.S., Dunshea, F.R., Ludbey, J., Lutz, A., & Bennett, L. (2018). Statistical modelling coupled with LC-MS analysis to predict human upper intestinal absorption of phytochemical mixtures. *Food Chemistry, 245*, 353–363.
42. De Mejia, E.G., & Dia, V.P. (2010). The role of nutraceutical proteins and peptides in apoptosis, angiogenesis, and metastasis of cancer cells. *Cancer and Metastasis Reviews, 29*(3), 511–528.
43. Leung, E.H., & Ng, T. (2007). A relatively stable antifungal peptide from buckwheat seeds with antiproliferative activity toward cancer cells. *Journal of Peptide Science: An Official Publication of the European Peptide Society, 13*(11), 762–767.

44. Guo, X., Zhu, K., Zhang, H., & Yao, H. (2010). Anti-tumor activity of a novel protein obtained from tartary buckwheat. *International Journal of Molecular Sciences, 11*(12), 5201–5211.
45. Shen, M., Chapman, R.S., He, X., Liu, L.Z., Lai, H., Chen, W., & Lan, Q. (2008). Dietary factors, food contamination and lung cancer risk in Xuanwei, China. Lung Cancer, *61*(3), 275–282.
46. Vogrinčič, M., Kreft, I., Filipič, M., & Žegura, B. (2013). Antigenotoxic effect of Tartary (Fagopyrum tataricum) and common (Fagopyrum esculentum) buckwheat flour. *Journal of Medicinal Food, 16*(10), 944–952.
47. Gimenez-Bastida, J.A., & Zielinski, H. (2015). Buckwheat as a functional food and its effects on health. *Journal of Agricultural and Food Chemistry, 63*(36), 7896–7913.
48. Li, F., Zhang, X., Zheng, S., Lu, K., Zhao, G., & Ming, J. (2016). The composition, antioxidant and antiproliferative capacities of phenolic compounds extracted from tartary buckwheat bran [Fagopyrum tartaricum (L.) Gaerth]. *Journal of Functional Foods, 22*, 145–155.
49. Li, W., Zhang, X., He, X., Li, F., Zhao, J., Yin, R., & Ming, J. (2020). Effects of steam explosion pretreatment on the composition and biological activities of tartary buckwheat bran phenolics. *Food & Function, 11*(5), 4648–4658.
50. Dzah, C.S., Duan, Y., Zhang, H., Boateng, N.A.S., & Ma, H. (2020). Latest developments in polyphenol recovery and purification from plant by-products: A review. *Trends in Food Science & Technology, 99*, 375–388.
51. Jeong, H., Sung, J., Yang, J., Kim, Y., Jeong, H.S., & Lee, J. (2018). Effect of sucrose on the functional composition and antioxidant capacity of buckwheat (Fagopyrum esculentum M.) sprouts. *Journal of Functional Foods, 43*, 70–76.
52. Lee, S.H., Lee, E.J., Min, K.H., Hur, G.Y., Lee, S.H., Lee, S.Y., Kim, J.H., Shin, C., Shim, J.J., & In, K.H. (2015). Quercetin enhances chemosensitivity to gemcitabine in lung cancer cells by inhibiting heat shock protein 70 expression. *Clinical Lung Cancer, 16*(6), e235–e243.
53. Wu, S.-C., & Lee, B.-H. (2011). Buckwheat polysaccharide exerts antiproliferative effects in THP-1 human leukemia cells by inducing differentiation. *Journal of Medicinal Food, 14*(1–2), 26–33.
54. Guo, X., Zhu, K., Zhang, H., & Yao, H. (2007). Purification and characterization of the antitumor protein from Chinese tartary buckwheat (Fagopyrum tataricum Gaertn.) water-soluble extracts. *Journal of Agricultural and Food Chemistry, 55*(17), 6958–6961.
55. Kayashita, J., Shimaoka, I., Nakajoh, M., Kishida, N., & Kato, N. (1999). Consumption of a buckwheat protein extract retards 7, 12-dimethylbenz [α] anthracene-induced mammary carcinogenesis in rats. Bioscience, Biotechnology, and Biochemistry, *63*(10), 1837–1839.
56. Zhang, L.-L., He, Y., Sheng, F., Hu, Y.-F., Song, Y., Li, W., Chen, J., Zhang, J., & Zou, L. (2021). Towards a better understanding of Fagopyrum dibotrys: A systematic review. *Chinese Medicine, 16*(1), 1–18.
57. Zhao, T., & Li, S. (2009). Extracts of Fagopyrum cymosum and Rosa roxburghii Tratt inhibit proliferation and induce apoptosis of human gastric and pulmonary carcinoma cells. *Journal of Practical Oncology, 24*(6), 542–547.
58. Liu, F.-d., Lv, Y.-q., Ye, K.-h., Zhao, J., & Ye, C. (2009). Effect of FR/MA on invasion of HepG2 human hepatocellular liver carcinoma cell line in vitro. *Journal of Practical Oncology, 24*, 337–340.
59. Mondal, S., Bhar, K., Sahoo, S.K., Seru, G., Ashfaquddin, M., Pradhan, N.K., Anjum, M., & Molla, S. Silver hull buckwheat (Fagopyrum esculentum Moench) is a part of nature that offers best health and honour. *Journal of Complementary and Alternative Medical Research*, *15*(2), 22–52.

60. Rodríguez-García, C., Sánchez-Quesada, C., & Gaforio, J.J. (2019). Dietary flavonoids as cancer chemopreventive agents: An updated review of human studies. *Antioxidants, 8*(5), 137.
61. Sabreena Nazir, M., Mahajan, R., Hashim, M.J., Iqbal, J., Alyemeni, M.N., Ganai, B.A., & Zargar, S.M. (2021). Deciphering allelic variability and population structure in buckwheat: An analogy between the efficiency of ISSR and SSR markers. *Saudi Journal of Biological Sciences, 28*, 11.

5 Cancer Prevention and Control
A Panoramic View

Mudasir Bashir
Graphic Era Institute of Medical Sciences (GEIMS), Dehradun, Uttrakhand

Moien Rasheed Lone
University of California, Los Angeles, California

Sheikh Imran Sayeed
Government Medical College Srinagar, Srinagar, India

5.1 INTRODUCTION

Cancer imposes a considerable strain on societies globally in terms of both financial burdens and epidemiological challenges [1,2]. Yearly, the majority of cancer diagnoses contribute to escalating mortality rates, positioning cancer as the second leading cause of death in the USA alone [3]. A spectrum of preventive measures has emerged, ranging from lifestyle modifications aimed at reducing risk to screening protocols designed to detect early-stage lesions along with proactive interventions to impede carcinogenic pathways. However, the landscape of cancer prevention is still evolving, with therapeutic advancements lagging behind those seen in diseases like cardiovascular conditions, where preventive strategies are more established. Despite this, recent updates underscore a growing emphasis on precision medicine and targeted interventions, promising avenues that could revolutionize cancer prevention efforts. As research progresses, optimizing these approaches will be pivotal in mitigating the impact of cancer on global health.

The intricate and diverse nature of cancer, in comparison with cardiovascular disorders, contributes to the slower pace of therapeutic development in the field. Cancer's complex biological landscape and poorly understood causal pathways present formidable challenges. Unlike cardiovascular diseases, reliable biomarkers for predicting cancer susceptibility are scarce, and our comprehension of factors influencing treatment response remains incomplete. Furthermore, the elusive nature of cancer pathways underscores the need for deeper exploration and elucidation. Enhancing our understanding of these intricate mechanisms holds the key to unlocking more effective preventive and therapeutic strategies. As research endeavors continue to unravel the complexities of cancer biology, strides toward personalized and precision

DOI: 10.1201/9781003477792-5

medicine are expected to accelerate, offering renewed hope in the battle against this multifaceted disease.

The recent epidemiological studies on cancer paint a concerning picture, with the World Health Organization predicting a significant increase in cancer prevalence among the general population. These studies project a steady rise in new cases annually, from 14 million in 2025 to 19 million in 2030 and a further increase to 21.7 million in 2035, reaching 24 million [4,5]. Globally conducted research emphasizes the effectiveness of adopting a healthy lifestyle, minimizing exposure to carcinogens, and undergoing routine screenings as the most reliable methods to reduce the risk of cancer [6]. Addressing the global challenge of increasing cancer prevalence necessitates a heightened focus on more potent preventive strategies, with both primary prevention and secondary prevention gaining prominence. These approaches offer numerous benefits, particularly considering cancer's lengthy latency period, which allows for the mitigation of causes and risk factors over time, significantly impacting public health. Interventions span a spectrum of initiatives to instigate behavioral changes and foster ongoing positive lifestyle modifications, extending beyond traditional medical or surgical procedures. A positive cycle can be initiated by fostering inspiration, education, and self-care practices, reducing the need for repetitive preventive measures across generations, especially crucial during periods of economic constraint [1].

Furthermore, regulatory measures can be pivotal in preventing various malignancies and associated pathologies. For instance, smoking, known to increase the risk of lung cancer, also elevates the likelihood of breast cancer [7], prostate cancer [8], lymphoma [9], and other diseases [10]. Leveraging the current understanding of etiopathogenesis and risk factors suggests that approximately half of all malignancies could be preventable [1]. Hence, connecting research with clinical practice through preventive medicine offers a swift and effective means of addressing this pressing issue.

5.2 OBJECTIVE

This review aims to consolidate insights into cancer prevention and control by analyzing both recent literature and historical literature relevant to the topic. We strive to provide a comprehensive understanding of cancer prevention and control by examining papers spanning various periods.

5.3 MATERIALS AND METHODS

The targeted literature search was conducted using the PubMed database, focusing on the published literature related to cancer prevention. Only those articles deemed most pertinent to the topic were selected from the plethora of articles retrieved. All chosen articles were written in English, and their abstracts were scrutinized to extract the most relevant data for this review. Furthermore, the full texts of potentially relevant articles were meticulously evaluated, particularly those elucidating the risk factors for cancer. This comprehensive approach was undertaken to enhance the presentation of study characteristics and results, facilitating readers to generate interpretations and conclusions through a thorough analysis of the existing literature.

5.4 RESULTS

Preventing cancer involves both primary and secondary measures. Primary prevention entails removing cancer-causing substances from the environment and dietary components linked to higher risk, as well as considering dietary supplements with potentially protective properties. Secondary prevention involves identifying and removing benign tumors and detecting breast cancer early through methods like mammography.

5.4.1 Avoiding Risky Exposures

Epidemiological studies strongly suggest that many cancers can be prevented. Cancer incidence varies among different populations and changes over time. For instance, migrants to Australia often adopt the cancer rates of their new home relatively quickly, while in some cases, like Japanese migrants to the USA, it may take generations. Additionally, subgroups within communities, such as Seventh day Adventists, Mormons, or certain racial groups in the USA, exhibit distinct cancer patterns. It's widely believed that 80%–90% of cancers in humans are influenced by environmental factors, broadly defined to encompass various lifestyle choices, dietary habits, social factors, and cultural practices. Despite this understanding, identifying and mitigating the avoidable causes of many common cancers remains a challenge.

5.4.2 Tobacco

Tobacco smoking remains the leading preventable cause of premature death worldwide and is the most significant known carcinogen for humans [11]. In developed nations, about 25%–30% of all cancers are linked to tobacco use. For instance, in 1990, tobacco smoking was responsible for three-quarters of a million middle-aged deaths (between 35 and 69 years) in Europe alone. Specifically, within the European Union member states during that year, over a quarter of a million middle-aged deaths were directly attributed to tobacco smoking, with 219,700 in men and 31,900 in women. Moreover, there were numerous additional deaths attributable to tobacco among older age groups. In Central and Eastern European countries, including the former Union of Soviet Socialist Republics (USSR), there were 441,200 middle-aged deaths in men and 42,100 in women due to tobacco-related causes [12].

Based on 30 epidemiological studies, the US Environmental Protection Agency has determined that environmental tobacco smoke, also known as passive smoking, is a carcinogen for the human lungs. In the USA alone, it's estimated that tobacco use, predominantly from cigarette smoking, leads to approximately 434,000 deaths annually, with 112,000 of these deaths attributed to lung cancer [13-14]. Additionally, passive smoking contributes to an estimated 1,500 deaths in non-smoking women and 500 deaths in non-smoking men annually.

The global situation for women is concerning, with a growing number of female smokers and a significant rise in lung cancer cases among women, paralleling trends seen in men in many countries [15-16]. For instance, in Japan, since 1950, lung cancer mortality has surged tenfold in men and eightfold in women. China is experiencing a substantial burden of tobacco-related diseases, as the rise in cigarette smoking prevalence during the 1950s and 1960s is now translating into increased incidence and mortality from smoking-related cancers. In Central and Eastern Europe, tobacco smoking is responsible for over 400,000 premature deaths each year [11].

A comprehensive tobacco control program endorsed by organizations like the International Union Against Cancer, the World Health Organization, and the Europe Against Cancer Programme of the European Community focuses on several key strategies. Among these, prioritizing tobacco control is paramount in any cancer control initiative, as it's expected to yield the most significant impact on reducing both cancer incidence and mortality compared to other known strategies. Recommendations for tobacco control from the European Cancer Experts Committee [17] warrant active support from the global scientific community.

5.4.3 Ethanol Drinking

Alcohol consumption is linked to cancer in multiple areas, primarily affecting the oral cavity, pharynx, larynx, and esophagus [18]. The predominant effect appears to synergize with cigarette smoking. It's recommended that heavy alcohol consumers consider moderating their intake and quitting smoking. This health advice carries a significant potential impact, particularly considering that upper respiratory tract cancers contribute to over 850,000 new cases annually worldwide [19].

5.4.4 Sunlight Exposure

Although receiving significant attention in many developed nations, malignant melanoma, which saw an estimated 92,000 new cases in 1985, constitutes only 1.2% of the overall cancer burden [19]. The crucial takeaway is the importance of minimizing sun exposure, particularly by avoiding sunburns at all costs and being especially vigilant in protecting children [20].

5.4.5 Cancer-Causing Viruses

Strong evidence links the human papilloma virus to a higher risk of cervical cancer in women [21], the Epstein-Barr virus to nasopharyngeal cancer [22], the hepatitis B virus to primary liver cancer [23], and the human immunodeficiency virus (HIV) to Kaposi's sarcoma and certain lymphomas [24-25]. It's estimated that there are at least 850,000 new cases of cancer annually, where infectious processes play a significant role, accounting for approximately 11% of the global cancer burden. Implementing effective vaccination programs targeting these viral infections would yield significant benefits. The most substantial impact would be observed in developing regions, which account for over three-quarters of global liver cancer cases. By integrating hepatitis B virus vaccination into the primary immunization schedule for infants, approximately two-thirds of all cases in developed and developing countries could be prevented due to the timing of infection [26].

5.4.6 Chemoprevention

Researchers are exploring various classes of chemopreventive agents, both natural and pharmaceutical, to evaluate their effectiveness in preventing cancer [27]. They're also investigating different stages of abnormal cell development, ranging from early-stage preneoplasia to late-stage preneoplasia, to understand how these cancer preventive treatments impact cell proliferation and differentiation. Current chemopreventive approaches are categorized based on their mechanism of action, with a focus on identifying intermediate endpoints to assess their effectiveness in preventing cancer (see Table 5.1).

5.4.7 Dietary Regimens

Human epidemiological studies have primarily focused on the connections between dietary factors like fat and vitamin intake and cancer risk. While the summary of these findings may provoke debate, current evidence suggests that higher fat and red meat consumption is linked to increased risks of colorectal and likely prostate cancers. Conversely, fruit and vegetable rich diets reduce the risk of several cancers, such as lung, oral cavity, pancreatic, laryngeal, esophageal, bladder, and stomach cancers. Notably, there's less strong evidence of a significant association between fruit and vegetable intake and hormonally related cancers such as prostate, breast, ovarian, and endometrial cancers [28]. This relationship seems to be a widespread effect observed across various groups of fruits and vegetables, likely influenced by multiple diet components and various candidate mechanisms.

Numerous randomized trials have explored the impact of increased fruit and vegetable consumption, with ongoing efforts to pinpoint the specific molecules responsible for their apparent protective effects [29-32]. One significant finding emerged from a randomized intervention trial in China, where a reduction in mortality was observed. This discovery offers hope for the potential success of future chemoprevention strategies [33]. It's worth noting that the Chinese population studied had high cancer rates and a history of low dietary intake of several essential micronutrients. However, past major studies' failure to demonstrate any beneficial effects of β-carotene supplementation on cancer risk in populations with normal intakes raises important strategic questions for future directions in chemoprevention research.

Cancer prevention programs focusing on dietary factors can be broadly categorized into two main approaches: analyses of the effects of entire diets [29-30] and chemopreventive strategies targeting individual dietary components [25-26,34-37]. Whole dietary regimens aimed at inhibiting tumors typically involve reducing dietary fat intake, boosting fiber, fruit, and vegetable consumption [29-30,37-42], and administering specific micronutrients, including various naturally occurring nutrients.

Studies investigating low-fat diets include the Women's Health Trial Vanguard Study, which assessed the feasibility of reducing fat intake in postmenopausal women [41-42]. Additionally, the National Institute of Health's Women's Health Initiative explored the effects of a modified diet, calcium and vitamin D supplementation, and hormonal replacement therapy on breast and colon cancers, coronary heart disease, and osteoporosis in postmenopausal women. Furthermore, the Women's Intervention

TABLE 5.1
Chemopreventive Approaches along with Their Postulated Mechanisms and Associated Biomarkers

Chemopreventive Strategies [27]

Chemopreventive Approaches	**Approach Products**	**Postulated Mechanism**	**Biomarkers**
Avoid cancer-related agents in the environment	Tobacco products, high dietary fat, and high calories Chemicals (e.g., aflatoxin and heterocyclic amines) Viruses (e.g., hepatitis)	Reduce DNA damage/promotion	Reduced urinary and blood markers that pinpoint specific agents (DNA adducts, oxidation products, and alkylation products)
Antioxidants	Tocopherols, ascorbate, carotenes, and selenium β-Carotene	Reduce the risk of excessive oxidative DNA damage and cell toxicity Reduce oxidation reactions	Reduced DNA oxidative metabolites in blood and urine and proliferation Reduced hyperproliferation and modulated cell differentiation
Inhibit arachidonic acid or oxidation	NSAID plant phenolic compounds, flavonoids, conjugated linoleic acid, ώ-3 (n3) fatty acids, and cyclo-oxygenase-2 inhibitors	Inhibit arachidonic acid metabolism and oxidative reactions	Reduced lipoxygenase activity, cyclo-oxygenase activity, inflammation, and proliferation
Alter enzyme action	Isothiocyanates, Dithiolethiones, flavones, indoles, and sulfur compounds Indole-3-carbinol d-Glucarates	Alter enzyme detoxification Increase 2-OH-estradiol oxidation Inhibit glucuronidase	Modified carcinogenic metabolites in biological fluids and urine Increased urinary 2-OH/16-OH estradiol ratio Increased urinary glucuronide excretion (e.g., estrogens)
Calcium and vitamin D		Reduce bile acid, fatty acid irritation in the colon, and proliferation	Modified gene expression, reduced hyperproliferation and cytotoxicity, and increased serum 25-OH-vitamin D_3

(Continued)

TABLE 5.1 (*Continued*)
Chemopreventive Approaches along with Their Postulated Mechanisms and Associated Biomarkers

Chemopreventive Strategies [27]

Chemopreventive Approaches Approach Products	Postulated Mechanism	Biomarkers
Difluoromethylornithine	Inhibit ODC	Reduced ODC levels associated with cell proliferation
Dehydroepiandrosterone	Reduce G-6PDH activity	Reduced carcinogen activation and cell proliferation
Retinoids	Induce cell differentiation	Reversed squamous metaplasia and modulated cell differentiation markers and leukoplakia
Terpenes	Reduce oncogene isoprenylation and ubiquinone synthesis	Modulated HMG-CoA metabolic pathways and products
Anti-estrogens (e.g., tamoxifen)	Reduce estrogen-stimulated proliferation	Modulated proliferation, estrogen metabolism, and gene expression

NSAID, non-steroidal anti-inflammatory agent; ODC, ornithine decarboxylase; G-6PDH, glucose-6-phosphate dehydrogenase; HMG-COA, hydroxy-3-methylglutaryl coenzyme A.

Nutrition Study examined dietary fat reduction in women with previous breast cancer. At the same time, the Polyp Prevention Trial focused on low-fat diets with high fiber, fruit, and vegetable intake to prevent polyps [29].

Numerous trials have investigated the tumor-inhibiting effects of individual nutrients and pharmaceutical compounds (see Table 5.2). Among the larger studies are the Linxian study focusing on esophageal and stomach cancers [33], the α-Tocopherol, β-Carotene Cancer Prevention (ATBC) Study targeting lung cancer prevention [24], the β-Carotene and Retinol Efficacy Trial (CARET) for lung cancer [43], and the Physicians' Health Study focusing on cancer prevention and cardiovascular mortality reduction [44].

5.4.8 Cancer Prevention Trials

Numerous cancer prevention trials have been conducted or are currently ongoing to evaluate intervention strategies. These trials target pre-neoplastic stages and established neoplasms involving the general population or individuals at heightened risk. High-risk groups encompass those with significant exposure to risk factors and

TABLE 5.2
Clinical Trials with Nutritional Supplements and Modified Diets

Nutritional Supplements and Modified Diets	Organ Site(s)
Calcium	Breast, colon, and esophagus
β-Carotene and other carotenoids	Breast, cervix, colon, esophagus, lung, oral, skin, and stomach
Dietary modifications	Breast, colon, and skin
Indole-3-carbino	Cervix and larynx
Selenium	Liver, lung, and skin
Vitamin A	Lung, oral, and skin
Vitamin C	Colon and stomach
Vitamin D3 (and analogs)	Colon and breast
Vitamin E	Breast, colon, esophagus, oral, and prostate
Vitamins, combined	Cervix, colon, esophagus, lung, oral, and stomach
Vitamins and minerals combined	Colon, esophagus, lung, and oral

These clinical trials represent pioneering efforts to uncover the potential benefits of dietary interventions.

TABLE 5.3
Geographic Distribution of Chemoprevention Intervention Trials

Country	Organ Site Studied
USA	Breast, cervix, colon, prostate, head and neck, lung, oropharynx, and skin
Europe	Breast, colon, cervix, head and neck, lung, and stomach
Australia	Colon and lung
Africa	Skin and liver
Canada	Breast, colon, and oropharynx
China	Esophagus, lung, liver, and oropharynx
India	Oral
South America	Stomach

Global landscape of chemoprevention intervention trials showcasing the countries where these groundbreaking studies have been conducted.

those presenting paraneoplastic or neoplastic lesions, presumed to have elevated susceptibility.

Significant studies have addressed cardiovascular disease prevention by integrating cancer-related intermediate endpoints or biomarkers (see Table 5.3). These investigations encompass diverse approaches such as smoking cessation, dietary fat modification, and specific chemopreventive agents. Cancer prevention intervention trials have been conducted globally [45-46], spanning regions including Africa, Australia, Canada, China, various European countries, South America, the USA, and nations formerly part of the USSR.

The usefulness of surrogate endpoint biomarkers in clinical trials depends on how closely they relate to cancer development. While late-stage markers like dysplasias are closely linked to cancer progression, early-stage markers such as cell proliferation offer valuable insights into how chemopreventive agents work. Intermediate endpoints that have been studied include late-stage precancerous lesions (e.g., breast and cervix), recurrence of precancerous lesions (e.g., colon), presence of micronuclei (e.g., lung), cell proliferation (e.g., colon), and cancer development (e.g., breast and lung) [46].

The clinical trials have investigated various chemopreventive agents across different cancer types. These include vitamin A for lung, oral mucosa, and skin cancers; vitamin C for colon and stomach cancers; synthetic retinoids for breast, cervical, head and neck, lung, and skin cancers; beta-carotene and other carotenoids for breast, cervical, colon, lung, skin, and stomach cancers; fiber for colon cancer; selenium for liver, lung, and skin cancers; calcium for esophageal, bladder, and colon cancers; Non-Steroidal Anti-Inflammatory Drugs (NSAIDs) for bladder and colon cancers; tamoxifen for breast cancer; vaccination for liver cancer; and dietary regimens for breast, colon, and skin cancers [45-46].

Apart from primary and secondary prevention measures, as discussed above, lifestyle factors are estimated to contribute to over 80% of cancer cases. Hence, cancer prevention strategies emphasizing health promotion are paramount. This underscores the importance of raising awareness about the significance of healthy lifestyle choices, encompassing regular exercise, balanced nutrition, and other positive behaviors [47].

Cervical cancer holds the unfortunate distinction of being the most prevalent malignancy globally, ranking tenth in incidence. However, in terms of fatality, it ranks third after breast and colorectal cancers [48-50], although this trend is on the decline. Several significant risk factors contribute to the development of cervical cancer, including alcohol and tobacco use, unprotected sex, history of sexually transmitted diseases (STDs), early onset of sexual activity, low socioeconomic status, HIV infection, and immunosuppression [48-50]. As per the WHO recommendations, preventive vaccinations against genital forms of the human papillomavirus (HPV) represent a significant breakthrough in primary prevention, addressing both benign and malignant lesions associated with HPV [51].

In 2006, the European Medicines Evaluation Agency approved two vaccinations in Europe, demonstrating efficacy in lowering the risk of cervical dysplasia. Mass immunization of girls aged 12-15 [52-64] presents a significant opportunity to prevent cervical cancer. Additionally, mass screening serves as a secondary preventive measure to detect the disease early in asymptomatic individuals and reduce associated mortality rates. Screening techniques encompass cervical cytology (Pap smear or Papanicolaou), colposcopy, and HPV DNA testing [48-50,65].

As per the 2010 European Guidelines for Quality Assurance in Cervical Cancer Screening [66], along with recommendations from the American College of Obstetricians and Gynecologists (ACOG), the American Cancer Society (ACS), and the Centers for Disease Control and Prevention/Advisory Committee on Immunization Practices (CDC/ACIP), screening is advised for women aged 20-30. Women aged 21-29 should undergo Pap smears every 3 years, regardless of their sexual activity. If abnormal results occur, HPV DNA testing [66-67] is recommended. For women over 29, a combined HPV DNA test and Pap smear can be conducted every 5 years. Screening should generally cease between 60 and 65, barring abnormal findings [66-67].

Colorectal cancer ranks as the third most common cancer among men, following lung and prostate malignancies, and the second most prevalent among women, after breast cancer. It accounts for over 1,360,000 diagnosed cases annually, constituting approximately 10% of the total cancer burden. While slightly more prevalent in men, its incidence rises significantly after age 36. After lung, liver, and stomach cancers, colorectal cancer stands as the fourth leading cause of cancer-related deaths. Epidemiological studies have consistently highlighted a higher risk of this cancer among populations with diets rich in salt, refined sugar, red and processed meats, and animal fats. In line with these findings, the World Cancer Research Fund advises limiting red meat consumption to 500 g weekly.

The European Prospective Investigation into Cancer and Nutrition (EPIC) study revealed that individuals consuming 80 g or more of red meat daily faced a nearly one-third higher risk of developing colorectal cancer. This finding resonates with similar associations found in numerous articles linking red meat consumption to stomach and pancreatic cancers. Notably, red meat and sausages contain elevated heme levels, which can trigger inflammation and disrupt the mucosa of the large intestine, promoting the production of carcinogenic compounds by gut bacteria [68–70]. Furthermore, studies have highlighted the role of dietary folic acid deficiency in increasing the risk of colorectal, breast, and uterine cancers. Folic acid plays a crucial role in DNA methylation, nucleic acid synthesis, and S-adenosylmethionine production, thus exerting a preventive effect. Insufficient folate levels lead to DNA hypomethylation and chromosomal damage, fostering excessive cellular proliferation [71-73].

Additionally, epidemiological research underscores the link between alcohol consumption and cancer risk. Even modest alcohol intake has been associated with heightened cancer risk. For instance, a daily intake of two alcoholic beverages elevates the risk of colorectal cancer by 8%. In contrast, a daily intake of one alcoholic drink (equivalent to 6-8 g of pure ethanol) increases the risk of breast cancer by 11%. According to the American Institute for Cancer Research (AICR), the consumption of sugar or sugary beverages may not satisfy hunger like regular meals. These items often provide excess calories, which can contribute to weight gain, overweight, and obesity, all of which have been indirectly linked to cancer development. However, the direct link between sugar consumption and cancer development is considered weak [68].

Daily consumption of fruits and vegetables has significantly reduced the risk of colorectal, stomach, rectal, and breast cancers. Research from the EPIC study reveals that a daily intake of 400-800 g of fruits and vegetables can decrease the risk of stomach and lung cancers by a quarter and decrease the risk of mouth, throat, laryngeal, and esophageal cancers by up to a third [68]. Fruits and vegetables are abundant sources of beneficial nutrients such as carotenoids, ascorbic acid, and folic acid. They also contain bioactive compounds like flavonoids, phenols, isothiocyanates, indoles, and dietary fibers, all of which exhibit anti-carcinogenic properties [74-76].

Moreover, incorporating legumes, especially cooked beans, into one's diet can protect against colorectal cancer due to saponins, oligosaccharides, phenolic compounds, and isoflavonoids [77-78]. Various screening procedures are available for detecting colorectal cancer, with the most common being the stool test, also known as the fecal occult blood test (FOBT). Other variants of these examinations include the guaiac-based FOBT (gFOBT), stool DNA test (sDNA), and fecal immunohistochemical test (FIT) [79-80]. Screening for colorectal cancer typically begins at the age of 50, but individuals at a significantly higher risk should start screening earlier and more frequently. Guidelines recommend several screening methods for eligible subjects, including double-contrast barium enema (DCBE) every 5 years, annual FOBT, flexible sigmoidoscopy (FS) every 5-10 years (often combined with FOBT every 1-3 years), virtual colonoscopy every 5 years, and colonoscopy every 10 years [81].

Despite advancements in diagnosis and treatment, breast cancer remains a pressing public health issue [82], with an estimated 1,384,155 new cases diagnosed globally and approximately 459,000 resulting deaths [82]. Alarmingly, since 2008, both the incidence and associated mortality rates have increased by 18%. In the USA, the ACS reports that one in every eight women will develop breast cancer. Looking ahead, projections suggest that by 2050, there could be 3.2 million new cases of breast cancer diagnosed annually worldwide [82].

Risk factors for breast cancer encompass various aspects such as breast size [83], sedentary lifestyle, overweight and obesity [84-86], nulliparity and infertility, early onset of menstruation, first full-term pregnancy at age 30 or older, tobacco and alcohol consumption, hormone replacement therapy (HRT) involving estrogen and progestin, inherited genetic mutations (BRCA1 and BRCA2) [82,86,87], exposure to diethylstilbestrol (DES), and syndromes like Cowden and Li-Fraumeni syndromes [86,87]. However, protective factors include vegetable consumption and breastfeeding. In 2002, the International Agency for Research on Cancer (IARC) amassed sufficient data to establish a link between obesity and the development of several types of cancers, such as breast, esophageal, endometrial, colorectal, and kidney cancers. These findings were corroborated by the World Cancer Research Fund (WCRF), which affirmed strong evidence indicating that obesity heightens the risk of developing rectal, pancreatic, and gallbladder cancers. More recent data from 2014 and 2015 by the WCRF has also included ovarian and prostate cancers in the list of malignancies associated with obesity.

Recent studies have elucidated various mechanisms through which regular exercise can impact cancer risk. A sedentary lifestyle often leads to obesity due to insufficient physical activity, resulting in elevated levels of circulating estrogens, androgens, insulin, and insulin-like growth factors. These hormonal imbalances can influence both tumor development and the proliferation of cells. Moreover, a lack of physical activity extends the exposure of breast tissue to circulating ovarian hormones and contributes to food retention in the large intestine. This prolonged exposure potentially exacerbates the carcinogenic effects on the intestinal lining [88]. Research has firmly established that engaging in regular physical activity reduces the risk of breast, colorectal, and endometrial cancers. Furthermore, there is evidence suggesting that regular exercise may also decrease the risk of lung and prostate cancers.

Physical activity exerts a multifaceted impact on the body, influencing immune function by stimulating the production of natural killer cells by releasing catecholamines and cortisol.

Regular exercise also leads to the loss of fatty tissue, elevating adiponectin levels while reducing the levels of TNF-alpha and IL-6. This shift in adipokine levels can mitigate DNA disruption, inhibit apoptosis, and impede tumor invasion [89–92]. Physical exercise significantly influences cancer prevention. Following guidance from the AICR, engaging in regular physical activity not only reduces the risk of cancer but also helps prevent weight gain, overweight, and obesity, all of which are associated with heightened cancer risk. Researchers suggest that 150 minutes of brisk walking or 75 minutes of vigorous weekly exercise is sufficient for reducing cancer risk. However, greater physical activity offers even more advantages for cancer prevention and control. Engaging in moderate physical exercise for 45-60 minutes daily provides significant health benefits, and further benefits are conferred by more than 60 minutes of exercise daily [93-94].

The primary prevention of cancer, as discussed above, hinges on reducing exposure to both active and passive tobacco smoking, a known carcinogen. The WHO Framework Convention on Tobacco Control (FCTC) asserts that all tobacco products, whether smoked, chewed, or sniffed, contain various carcinogens and other hazardous substances. Carcinogenic compounds like nitrosamines (TSNAs) such as N-nitrosonornicotine (NNN) and 4-(methylnitrosamino)-1-(3-pyridyl)-1-butanone (NNK), as well as polycyclic aromatic hydrocarbons (PAHs) like benzo[a]pyrene [95-98], are present in tobacco leaves and are further generated upon combustion. The International Agency for Research on Cancer (IARC) identifies over 70 constituents of tobacco smoke as carcinogens, with 16 of them known to cause cancer in humans. Notable chemicals include benzene, dimethylnitrosamine, ethylmethyltitrosamine, diethylnitrosamine, nitrosopyrrolidine, hydrazine, and vinyl chloride [88,95]. Smoking, as confirmed by the ACS, is associated with at least 15 different types of cancers: hematopoietic, oral cavity, laryngeal, throat, large intestine, lung, nasal cavity, paranasal sinus, stomach, esophageal, liver, pancreatic, bladder, cervical, and breast cancers. Moreover, evidence links smoking to childhood leukemia and breast cancer [88,95,99]. Passive smoking poses an equal risk of cancer compared to active smoking. The smoke emitted from the burning end of a cigarette is four times more toxic than the smoke directly inhaled by the smoker. According to the American Cancer Society Report, exposure to secondhand smoke increases the risk of lung cancer by a significant quarter. Furthermore, it elevates the likelihood of developing various cancers, such as esophageal, laryngeal, brain, throat, bladder, rectal, and stomach cancers [95,100-101]. Additionally, it is associated with pediatric leukemia.

Infectious agents play a significant role in the complex landscape of cancer development. Therefore, minimizing exposure to biological agents such as bacteria, viruses, fungi, and protozoa is crucial. Preventive strategies encompass maintaining optimal housing conditions, ensuring environments are dry and free from mold, and mitigating the risk of infection, particularly in infants under three months old. For individuals with a familial history of liver disease or cancer, proactive screening for hepatitis B virus (HBV) and hepatitis C virus (HCV) is recommended. Additionally, infants born to mothers carrying HBV or HCV should receive immunoglobulins as a preventive measure. Moreover, carriers of human T-lymphotropic virus type 1 (HTLV-1) are advised against breastfeeding to mitigate transmission risks to their offspring. Individuals with a family history of stomach cancer are encouraged to undergo testing for Helicobacter pylori. Furthermore, protective immunizations are pivotal in primary prevention efforts. These vaccinations should encompass mandated ones and supplemental immunizations targeting infections like hepatitis A and hepatitis B, Haemophilus influenzae type B, chickenpox, mumps, influenza, and pneumococcal infections [102].

5.5 NATURAL COMPOUNDS IN CANCER PREVENTION

Cancer prevention and therapy pose formidable challenges, driving extensive efforts to both prevent the onset and progression of cancer and treat oncologic patients pharmacologically. Despite these efforts, many types of cancers lack effective therapeutic strategies, and the side effects associated with current chemotherapy underscore the need for alternative approaches. Natural extracts offer a promising avenue for discovering new chemical entities that could serve as the foundation for novel anticancer drugs. Moreover, natural extracts offer significant benefits in cancer prevention, primarily through their antioxidant and anti-inflammatory properties. In this context, there is a growing interest in identifying and characterizing new compounds within mixtures and extracts and assessing their protective and therapeutic effects against cancer progression. This pursuit holds the potential for discovering novel treatments and contributes to understanding how natural compounds can impact cancer biology.

Natural products are chemical compounds or substances synthesized by living organisms. They have been utilized since ancient times, with historical records dating back to Mesopotamia around 2600 BC. The "Ebers Papyrus," documenting over 700 drugs, provides evidence of their early use around 1550 BC. Traditional Chinese medicine and the Indian Ayurvedic system have been employed since the first millennium BC [103].

The precise mechanisms underlying the anticancer effects of natural products remain elusive, but several actions are believed to contribute to their efficacy. These include targeting cancer at various stages of development, influencing both extrinsic and intrinsic apoptosis pathways, modulating autophagic pathways and embryonic developmental pathways, bolstering the immune system, and inhibiting cancer cell-related activating proteins, various DNA enzymes, and signaling pathways. They also target cancer-associated pathways such as Hippo signaling, Wnt/β-catenin, Janus kinase (JAK)/signal transducer and activator of transcription (STAT), phosphoinositide 3-kinase (PI3K)/Akt/mammalian target of rapamycin (mTOR), hypoxia-inducible factor-1α (HIF-1α), and activator protein 1 (AP-1). Moreover, natural products can improve the local hypoxia of solid tumors by enhancing vascular permeability and augmenting the effects of radiotherapy. These mechanisms may involve enhancing hormonal regulation, promoting pituitary-adrenocortical function, and increasing the relative value of cyclic adenosine monophosphate. Table 5.4 summarizes some of the natural compounds/products along with their source and mode of action that are being used therapeutically in cancer treatment and prevention.

TABLE 5.4
List of Some of the Natural Product Compounds Utilized in Cancer Therapy, Detailing Their Botanical or Biological Origins and Elucidating Their Mechanisms of Action

Chemical Compounds	Structure	Major Source	Mechanism	Remarks	Ref
3,3′-Diindolylmethane	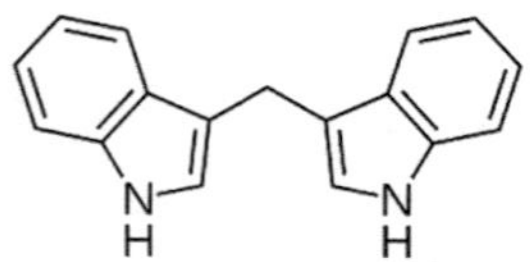	Broccoli Cabbage Cauliflower	Arachidonic acid pathway	Reduces the expression level of COX-2 in MCF-7 breast cancer cells	[104]
			Apoptosis pathway	Survivin, cdc25A, and Bcl2 downregulation p21 (WAF1) overexpression	[105]
			HIF-1 signaling pathway	Reduces the expression levels of furin, VEGF, glucose transporter-1, and enolase-1 in hypoxic tumor cell lines	[106]
			Aromatase activity	Reduces the expression level of aromatase	[107]
			Chemosensitivity/adjuvant therapy	Lowers the expression levels of FOXM1, Akt, and NF-κB p65, boosting the effectiveness of Herceptin Increases tamoxifen's chemosensitivity and modifies the estrogen metabolism	[108]

(*Continued*)

TABLE 5.4 (*Continued*)

List of Some of the Natural Product Compounds Utilized in Cancer Therapy, Detailing Their Botanical or Biological Origins and Elucidating Their Mechanisms of Action

Chemical Compounds	Structure	Major Source	Mechanism	Remarks	Ref
				Stimulates MDR human breast cancer to gamma irradiation	
Biochanin A	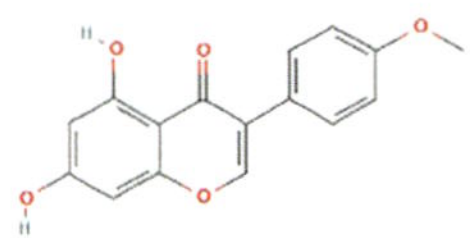	Red clover	Aromatase activity Cytoplasmic signaling pathway	Impedes the expression of aromatase	[109–111]
				Suppresses the activation of the HER-2 receptor Suppresses Erk1/2, Akt, mTOR, NF-κB, MMP-9, and MT-MMP1 With a dose of 15 mg/kg, inhibits the development of tumors in a xenograft animal model	
Curcumin	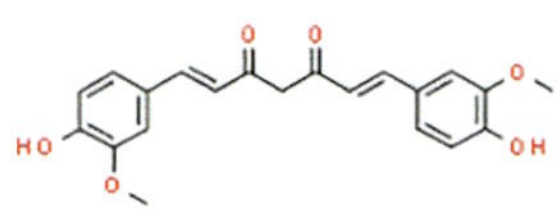	Turmeric	Apoptosis pathway Wnt signaling Pathway Epigenetic regulation Chemosensitivity/ adjuvant therapy	Increases p53 level	[112–117]

(*Continued*)

TABLE 5.4 (*Continued*)
List of Some of the Natural Product Compounds Utilized in Cancer Therapy, Detailing Their Botanical or Biological Origins and Elucidating Their Mechanisms of Action

Chemical Compounds	Structure	Major Source	Mechanism	Remarks	Ref
				Bax overexpression	
				Transcription of NF-kB, cyclin D, and MMP-1 is downregulated in the MDA-MB-231 and BT-483 cell lines	
				5 µM dosage blocks Wnt signaling in MCF-7 cell line	
				Suppresses the expression of class I HDACs	
				Induces the expression of certain miRNAs to limit Bcl-2 expression	
				Boosts the efficacy of paclitaxel by inhibiting the expression of MMP-9 and NF-B in MDA-MB-231 cells	

(*Continued*)

TABLE 5.4 (*Continued*)
List of Some of the Natural Product Compounds Utilized in Cancer Therapy, Detailing Their Botanical or Biological Origins and Elucidating Their Mechanisms of Action

Chemical Compounds	Structure	Major Source	Mechanism	Remarks	Ref
				Prevents P-gp, MXR, and MRP1 from performing their efflux function in the MCF-7 and MCF-7MDR cell line	
Lycopene	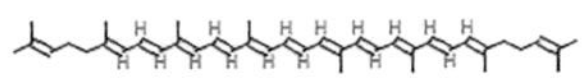	Tomato, carrot,watermelon,papaya, and cherry	Epigenetic regulation	Elevates GSTP1 expression	[118]
			Apoptosis pathway		[118,119]
				Demethylates GSTP1 promoter	
				Inhibits the Akt downstream pathway's phosphorylation	
				Elevates proapoptotic Bax expression without impacting antiapoptotic Bcl-xL	
				Reduces cyclin D1 and increases p21	
				Maintains the ERK1/2's state of activation	
				Increases p53 and Bax expression	

5.5.1 3,3′-Diindolylmethane (DIM)

This naturally occurring compound, abundant in cruciferous vegetables like cauliflower, broccoli, and cabbage, is a noteworthy acid condensation product of indole-3-carbinol (I3C) [120]. In the stomach's acidic environment, I3C converts to DIM [121]. Experimental evidence suggests that DIM reduces COX-2 expression mediated through the aryl hydrocarbon receptor in human breast cancer cells [104]. Additionally, Fan et al. demonstrated that DIM enhances BRCA1 phosphorylation under oxidative stress, acting as a preventive factor [122]. Previous research indicates that DIM restrains the expression of genes implicated in angiogenesis, such as survivin [105] and hypoxia-inducible factor-1α [106]. Furthermore, in HER-2/Neu-expressing breast cancer cells, DIM combined with Herceptin suppresses FOXM1 expression by downregulating Akt and NF-κBp65 [108]. Additionally, Wang et al. found that DIM enhances γ-radiation effects, inducing apoptosis by arresting the cell cycle at the G2/M phase and increasing intracellular ROS levels [107]. Furthermore, DIM acts as an aromatase inhibitor by reducing aromatase expression in MCF-7 cells and increasing CYP19 expression in MDA-MB-231 cells [123]. Recent research by Marques et al. [124] revealed that at concentrations of 10 μM, DIM stimulates estrogen receptor-α signaling pathways, leading to cell proliferation in the absence of estradiol, despite its significant impact on breast cancer treatment. This finding suggests a potential risk associated with DIM dietary supplementation. Thus, further investigation is necessary to fully understand the unexpected adverse effects of DIM supplementation for breast cancer treatment or prevention.

5.5.2 Biochanin A

Biochanin A, an isoflavone found in red clover (Trifolium pratense), exhibits promising anticancer properties [125]. Wang et al. [109] demonstrated that biochanin A effectively inhibits aromatase enzyme activity, thereby impeding cell proliferation in MCF-7 cells transfected with the CYP19 gene. Moreover, this compound suppresses aromatase enzyme activity and decreases mRNA expression in SK-BR3 cells, which are estrogen receptor (ER)-negative breast cancer cells. Genistein, a metabolite of biochanin A, has demonstrated the ability to inhibit promoter I.3/II activation, acting as an aromatase inhibitor (AI). Previous research suggests that biochanin is better tolerated compared to genistein, and it promotes the positive expression of tumor suppressor genes in various cell lines, including HMEC, MCF-12A, and MCF-7 (an ER-positive breast cancer cell line) [110, 126]. Young et al. likewise found that biochanin A exhibited superior tumor suppressor gene expression induction compared to genistein [127]. Bhushan et al. reported that biochanin A reduced the expression and activity of invasive enzymes and the viability and signaling of SK-BR-3 cancer cells [110]. In a xenograft mouse model, Moon et al. demonstrated that biochanin A, administered at doses of 5 or 15 mg/kg/day, effectively inhibited the growth of estrogen-dependent MCF-7 tumors [111]. Further investigation through clinical trials is essential to comprehensively grasp the bioavailability, optimal therapeutic dosage, and metabolic behavior of biochanin A across different types of breast cancer patients. Additionally, there is a need to elucidate the impact of biochanin A on other pathways involved in breast cancer progression.

5.5.3 Curcumin

Curcumin, the primary active compound in turmeric, is a polyphenolic substance widely recognized for its diverse medicinal applications, including breast cancer prevention. Various studies indicate that curcumin regulates the expression of genes and proteins associated with apoptosis, ultimately inducing the demise of breast cancer cells [128]. Recent research suggests that curcumin enhances p53 levels, thereby upregulating Bax expression and increasing the Bax/Bcl-2 ratio, triggering apoptosis in breast cancer cells. This cascade of events culminates in programmed cell death [129]. Research indicates that curcumin can inhibit the expression of NF-κB, a pivotal signaling protein involved in cell growth regulation. By downregulating NF-κB expression, curcumin exerts an antiproliferative effect on MDA-MB-231 and BT-483 cells [112]. Furthermore, another study suggests that curcumin can impede breast cancer metastasis by decreasing the protein expression of urokinase-type plasminogen activator through NF-κB activation. Curcumin can potentially hinder MCF-7 cells from adhering and becoming invasive [130]. Kakarala et al. discovered that curcumin effectively suppressed Wnt signaling in MCF-7 cells, a pathway often dysregulated in breast cancer [113,131-132]. This pathway is vital for breast stem cell self-renewal. Moreover, curcumin regulates the enzymatic activity of Histone Acetyltransferases (HATs) and Histone Deacetylases (HDACs), enabling it to modify histones.

Curcumin has been shown to inhibit the expression of class I HDACs effectively [114]. Moreover, it suppresses Bcl-2 expression by upregulating specific miRNAs implicated in carcinogenesis [115]. Additionally, curcumin enhances the efficacy of chemotherapy drugs. By inhibiting NF-κB expression, curcumin can augment the effectiveness of paclitaxel. This combined approach results in the inhibition of breast cancer growth in MDA-MB-231 (ER–/PR–) cells. When paclitaxel is paired with curcumin, there is a reduction in tumor size and growth, an increase in apoptosis rates, and a downregulation of MMP-9 expression [116]. Curcumin's ability to block this pathway positions it as a promising anti-breast cancer medication.

5.5.4 Lycopene

Lycopene, a phytochemical and tetraterpenoid, is a vibrant red carotene pigment naturally occurring in foods such as carrots, watermelons, tomatoes, papayas, and cherries. Functioning as a natural antioxidant, lycopene regulates numerous genes associated with DNA repair, cell cycle regulation, and apoptosis in breast cancer cells [117,133]. King-Batoon et al. demonstrated the impact of lycopene on the GSTP1 gene in breast cancer cells [134]. In the MDA-MB-468 cell line, lycopene treatment (2 μM for one week) was found to enhance GSTP1 expression and demethylase the GSTP1 promoter. However, the response differs in MCF-7 breast cancer cells. Treatment with lycopene in both MCF-7 and MDA-MB-468 breast cancer cells did not impact the expression of other genes, such as RARβ2 and HIN1 [134-135]. Furthermore, lycopene exerts its anticancer effects by modulating cell growth factor signaling pathways and inducing cell cycle arrest, ultimately leading to cell death. Takeshima et al. [118] conducted similar research, showing that lycopene may inhibit cell growth by blocking the activation of Akt through its downstream pathway. Additionally, it can elevate pro-apoptotic Bax levels without affecting anti-apoptotic Bcl-xL in triple-negative breast cancer cells.

Moreover, lycopene was found to upregulate p21 expression and prolong the activation of ERK1/2 while suppressing cyclin D1. Subsequently, Peng et al. unveiled lycopene's anti-proliferative role in MCF-7 cell lines, illustrating its ability to regulate the expression of p53 and Bax, thereby inhibiting cell proliferation and promoting apoptosis [119]. The role of lycopene in ER subtypes in breast cancer is still under discussion. One study suggested a potential negative association between lycopene and other carotenoids with ER-negative subtypes. At the same time, another study indicated that consuming carotenoids like lycopene might decrease the risk of developing ER-negative subtypes [136]. This highlights the need for more research to clarify how lycopene functions as a preventive measure in different pathways and across various breast cancer subtypes.

5.6 CONCLUSION

Research indicates that preventive programs play a pivotal role in reducing both the occurrence of cancer and the associated mortality rates. These programs are indispensable in the fight against cancer, particularly as screening tests continue to be underutilized, and adherence to screening programs remains a challenge that requires attention. Beyond the individual level, preventive measures constitute a vital aspect of health policy. The significance and impact of natural compounds in treating and preventing breast cancer cannot be overstated and should not be overlooked.

The utilization of dietary compounds derived from natural sources is widespread and has historical roots. There is a pressing need to prioritize research endeavors aimed at enhancing the efficacy of these compounds and harnessing them as a therapeutic approach for breast cancer patients. Additionally, education plays a pivotal role in prevention strategies, as it fosters health awareness and significantly influences health behaviors, such as adherence to scheduled testing at regular intervals [106-107] (Figure 5.1).

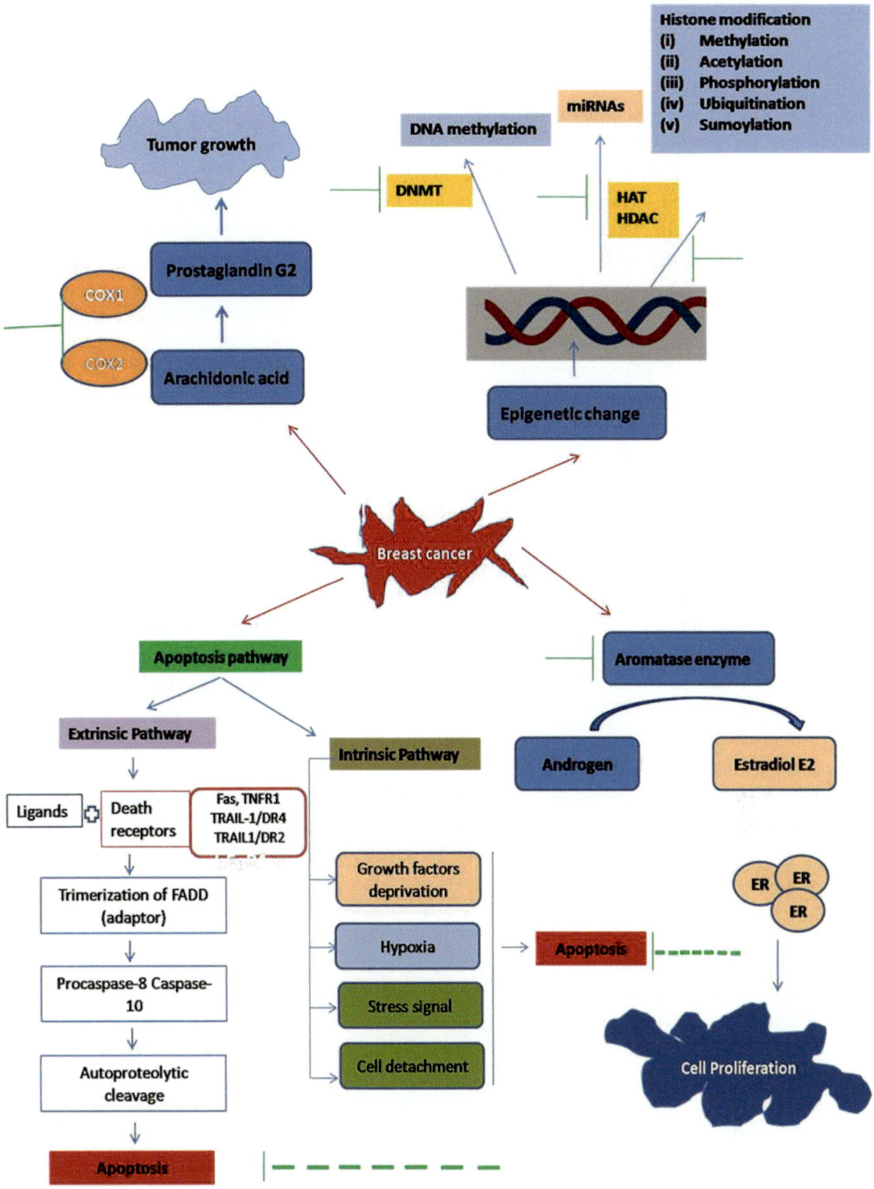

FIGURE 5.1 Molecular targets of natural compounds within the breast cancer pathway. In the figure, green solid lines depict the inhibitory action of natural compounds on specific molecular targets, while green dotted lines illustrate the upregulation induced by natural compounds. This figure was adopted from Sarmistha Mitra and Raju Dash et al. (137) and modified.

REFERENCES

1. Vineis P, Wild CP. Global cancer patterns: causes and prevention. *Lancet.* 2014;383:549–57.
2. Ehni HJ. Expensive cancer drugs and just health care. *Best Pract Res Clin Gastroenterol.* 2014 Apr;28(2):327–37.
3. American Cancer Society. *Cancer facts and figures.* Atlanta; 2015.
4. Wiszniewska M, Magnuska J, Lipińska-Ojrzanowska A, Pepłońska B, Hanke W, Kalinka J, et al. The prophylactic operational model integrated with occupational healthcare - Prophylactic of some types of cancers among women. *Med Pr.* 2018 Aug 20;69(4):439–55.
5. Global Burden of Disease Cancer Collaboration, Fitzmaurice C, Dicker D, Pain A, Hamavid H, Moradi-Lakeh M, et al. The global burden of cancer 2013. *JAMA Oncol.* 2015 Jul;1(4):505–27.
6. Adamowicz K, Zaucha JM, Majkowicz M. Assessment of the state of knowledge regarding cancer prevention among patients of the breast cancer prevention clinic Nowotwory. *J Oncol.* 2011;61(5):449–56.
7. Terry PD, Miller AB, Rohan TE. Cigarette smoking and breast cancer risk: a long latency period? *Int J Cancer.* 2002;100(6):723–8.
8. Islami F, Moreira DM, Boffetta P, Freedland SJ. A systematic review and meta-analysis of tobacco use and prostate cancer mortality and incidence in prospective cohort studies. *Eur Urol.* 2014;66(6):1054–64.
9. Kamper-Jørgensen M, Rostgaard K, Glaser SL, Zahm SH, Cozen W, Smedby KE, et al. Cigarette smoking and risk of Hodgkin lymphoma and its subtypes: a pooled analysis from the International Lymphoma Epidemiology Consortium (InterLymph). *Ann Oncol.* 2013;24(9):2245–55.
10. Samet JM. Tobacco smoking: the leading cause of preventable disease worldwide. *Thorac Surg Clin.* 2013;23(2):103–12.
11. Peto R, Lopez AD, Boreham J, Thun M, Heath C. Mortality from tobacco in developed countries: indirect estimation from national vital statistics. *Lancet.* 1992;39:1268–78.
12. Peto R, Lopez AD, Boreham J, Thun M, Heath C, Doll R. Mortality from smoking worldwide. *Br Med Bull.* 1996;52:12–21.
13. Centers for Disease Control. Smoking-attributable mortality and years of potential life lost–United States, 1988. *MMWR.* 1991;40:62–71.
14. USDHHS (United States Department of Health and Human Services). *Reducing the health consequences of smoking: 25 years of progress, a report of the Surgeon General.* Washington, DC: US Government Printing Office; 1989.
15. La Vecchia C, Lucchini F, Negri E, Boyle P, Maisonneuve P, Levi E. Trends of cancer mortality in Europe, 1955–1989: II respiratory tract, bone, connective and soft tissue sarcomas, and skin. *Eur J Cancer.* 1992;28:514–99.
16. La Vecchia C, Lucchini F, Negri E, Boyle P, Levi F. Trends in cancer mortality in the Americas, 1955–1989. *Eur J Cancer.* 1993;29A:431–70.
17. Boyle P. Cancer, cigarette smoking and premature death in Europe: a review including the recommendations of European Cancer Experts Consensus Meeting, Helsinki, October 1996. *Lung Cancer.* 1997 May;17(1):1–60.
18. IARC (International Agency for Research on Cancer). *Monographs on the evaluation of carcinogenic risk to humans: alcohol drinking.* Vol 44. Lyon: IARC; 1988.
19. Parkin DM, Ferlay J, Pisani P. Estimates of the worldwide incidence of eighteen major cancers in 1985. *Int J Cancer.* 1993;54:594–606.
20. Boyle P, Veronesi U, Tubiana M, Alexander FE, da Silva FC, Denis LJ, et al. European School of Oncology advisory report to the European Commission for the "Europe Against Cancer Programme": European code against cancer. *Eur J Cancer.* 1995;9:1395–405.

21. IARC (International Agency for Research on Cancer). *Monographs on the evaluation of carcinogens to man: human papilloma viruses.* Vol XX. Lyon: IARC; 1996.
22. Epstein MA. Epstein-Barr virus – discovery, properties and relationship to nasopharyngeal carcinoma. In: De Th G, Ito Y, eds. *Nasopharyngeal carcinoma: etiology and control.* Lyon: IARC, Scientific Publication No 20; 1978.
23. Beasley RP, Lin C, Hwang LY, Chien CS. Hepatocellular carcinoma and hepatitis B virus: a prospective study of 22 707 men in Taiwan. *Lancet.* 1981;ii:1129–33.
24. Biggar RJ, Rosenberg PS, Cote T. Kaposi's sarcoma and non-Hodgkin's lymphoma following the diagnosis of AIDS: multistate AIDS/cancer match study. *Int J Cancer.* 1996;68:754–58.
25. Pedersen C, Barton SE, Chiesi A, Skinhøj P, Katlama C, Johnson A, et al. HIV-related non-Hodgkin's lymphoma among European AIDS patients. AIDS in Europe Study Group. *Eur J Haemaotol.* 1995;55:245–50.
26. Tomatis L, Aitio A, Day NE. *Cancer: causes, occurrence and control.* Lyon: IARC, Scientific Publication No 100; 1990.
27. Lipkin M. Strategies for colon cancer prevention. *Ann NY Acad Sci.* 1995;768:129–40.
28. Steinmetz KA, Potter JD. Vegetable, fruit, and cancer I: epidemiology. *Cancer Causes Control.* 1991;2:325–58.
29. Greenwald P, Clifford C, Pilch S, Heimendinger J, Kelloff G. New directions in dietary studies in cancer: the National Cancer Institute. In: Longnecker JB, et al., eds. *Nutrition and biotechnology in heart disease and cancer.* New York: Plenum Press; 1995. pp. 229–39.
30. Weisberger JH. Nutritional approach to cancer prevention with emphasis on vitamins, antioxidants, and carotenoids. *Am J Clin Nutr.* 1995;53:226s.
31. Wattenberg L, Lipkin M, Boone CW, Kelloff G, eds. *Chemoprevention of cancer.* Boca Raton, FL: CRC Press; 1992.
32. Kelloff G, Boone C, eds. Cancer chemopreventive agents: drug development status and future prospects. *J Cell Biochem.* 1994;20:1–303.
33. Blot WJ, Li J-Y, Taylor P, Guo W, Dawsey S, Wang GQ, et al. Nutrition intervention trials in Linxian, China: supplementation with specific vitamin/mineral combinations, cancer incidence, and disease-specific mortality in the general population. *J Natl Cancer Inst.* 1993;85:1483–92.
34. The Alpha-Tocopherol, Beta-Carotene Cancer Prevention Study Group. The effect of vitamin E and beta carotene on the incidence of lung cancer and other cancers in male smokers. NEnglJMed 1994; 330: 1029–35.
35. Hennekens CH, Bnring ~E, Manson JE, et al. Lack of effect of longterm supplementation with beta-carotene on the incidence of malignant neoplasms and cardiovascular disease. N Engl.7 Med 1996; 334: 1145–49.
36. Omenn GS, Goodman GE, Thoruquist MD, et al. Effects of combination of beta-carotene and vitamin A on lung cancer and cardiovascular disease. NEnglJMed 1996; 334" 1150–55.
37. Buiatti E. Intervention trials of cancer prevention: results and new research programmes. Lyon: IARC Technical Report No 18; 1994.
38. Armstrong B, Doll R. Environmental factors and cancer incidence and mortality in different countries, with special reference to dietary practices. *Int J Cancer.* 1975;15:617–31.
39. Hursting S, Thornquist M, Henderson M. Types of dietary fat and the incidence of cancer at five sites. *Prey Med.* 1990;19:242–53.
40. Rose D, Goldman M, Connolly J, Strong L. High-fiber diet reduces serum estrogen concentrations in premenopausal women. *Am Clin Nutr.* 1991;54:520–25.
41. Henderson M, Kushi L, Thompson D, Gorbach SL, Clifford CK, Insull W, et al. Feasibility of a randomized trial of a low-fat diet for the prevention of breast cancer: dietary compliance in the Women's Health Trial vanguard study. *Prev Med.* 1990;19:115–33.

42. White T, Shattuck A, Kristal A, Urban N, Prentice RL, Henderson MM, et al. Maintenance of a low-fat diet: follow-up of the women's health trial. *Cancer Epidemiol Biomark.* 1992;1:315–22.
43. Omenn G, Goodman G, Thornquist M, Grizzle J, Rosenstock L, Barnhart S, et al. The beta-carotene and retinol efficacy trial (CARET) for chemoprevention of lung cancer in high risk populations; smokers and asbestos-exposed workers. *Cancer Res.* 1994;54:2038s–43s.
44. Hennekens C. Issues ha the design and conduct of clinical trials. *J Natl Cancer Inst.* 1984;73:1473–76.
45. Kelloff G, Boone C, eds. Cancer chemopreventive agents: drug development status and future prospects. *J Cell Biochem.* 1994;20:1–303.
46. Buiatti E. Intervention trials of cancer prevention: results and new research programmes. Lyon: IARC Technical Report No 18; 1994.
47. Carlos RC, Fendrick AM. Improving cancer screening adherence: using the "teachable moment" as a delivery setting for educational interventions. *Am J Manag Care.* 2004;10(4):247–8.
48. McGraw SL, Ferrante JM. Update on prevention and screening of cervical cancer. *World J Clin Oncol.* 2014;5(4):744.
49. Köse FM, Naki MM. Cervical premalignant lesions and their management. *J Turk Ger Gynecol Assoc.* 2014;15(2):109.
50. Aggarwal P. Cervical cancer: can it be prevented? *World J Clin Oncol.* 2014;5(4):775.
51. Lewandowska AM, Lewandowski T, Rudzki M, Rudzki S, Laskowska B. Cancer prevention–review paper. *Ann Agric Environ Med.* 2021;28(1):11–9.
52. Griesser H, Sander H, Walczak C, Hilfrich RA. HPV vaccine protein L1 predicts disease outcome of high-risk HPV+ early squamous dysplastic lesions. *Am J Clin Pathol.* 2009;132(6):840–5.
53. Carlos RC, Dempsey AF, Patel DA, Dalton VK. Cervical cancer prevention through human papillomavirus vaccination: using the "teachable moment" for educational interventions. *Obstet Gynecol.* 2010;115(4):834.
54. Khan K, Curtis CR, Ekwueme DU, Stokley S, Walker C, Roland K, et al. Preventing cervical cancer: overviews of the National Breast and Cervical Cancer Early Detection Program and 2 US immunization programs. *Cancer.* 2008;113(S10):3004–12.
55. Dempsey AF, Gebremariam A, Koutsky L, Manhart L. Behavior in early adolescence and risk of human papillomavirus infection as a young adult: results from a population-based study. *Pediatrics.* 2008;122(1):1–7.
56. Rand CM, Shone LP, Albertin C, Auinger P, Klein JD, Szilagyi PG. National health care visit patterns of adolescents: implications for delivery of new adolescent vaccines. *Arch Pediatr Adolesc Med.* 2007 Mar;161(3):252–9.
57. Insinga RP, Dasbach EJ, Elbasha EH. Assessing the annual economic burden of preventing and treating anogenital human papillomavirus-related disease in the US. *Pharmacoeconomics.* 2005;23(11):1107–22.
58. Physicians AA of F. Joint principles of the patient-centered medical home. *Del Med J.* 2008;80(1):21–2.
59. Blandin Knight S, Crosbie PA, Balata H, Chudziak J, Hussell T, Dive C. Progress and prospects of early detection in lung cancer. *Open Biol.* 2017;7(9):170070.
60. Smith RA, Andrews KS, Brooks D, Fedewa SA, Manassaram-Baptiste D, Saslow D, et al. Cancer screening in the United States, 2017: a review of current American Cancer Society guidelines and current issues in cancer screening. *CA Cancer J Clin.* 2017;67(2):100–21.
61. Oken MM, Hocking WG, Kvale PA, Andriole GL, Buys SS, Church TR, et al. Screening by chest radiograph and lung cancer mortality: the prostate, lung, colorectal, and ovarian (PLCO) randomized trial. *JAMA.* 2011 Nov 2;306(17):1865–73.

62. Franco EL, de Sanjosé S, Broker TR, Stanley MA, Chevarie-Davis M, Isidean SD, et al. Human papillomavirus and cancer prevention: gaps in knowledge and prospects for research, policy, and advocacy. *Vaccine*. 2012;30:F175–82.
63. Wentzensen N, Arbyn M. HPV-based cervical cancer screening-facts, fiction, and misperceptions. *Prev Med*. 2017;98:33–5.
64. Demarco M, Carter-Pokras O, Hyun N, Castle PE, He X, Dallal CM, et al. Validation of a human papillomavirus (HPV) DNA cervical screening test that provides expanded HPV typing. *J Clin Microbiol*. 2018;56(5):e01910–17.
65. Lee SH, Vigliotti JS, Vigliotti VS, Jones W. From human papillomavirus (HPV) detection to cervical cancer prevention in clinical practice. *Cancers*. 2014;6(4):2072–99.
66. Arbyn M, Anttila A, Jordan J, Ronco G, Schenck U, Segnan N, et al. European guidelines for quality assurance in cervical cancer screening. Second edition—summary document. *Ann Oncol*. 2010 Mar;21(3):448–58.
67. Haugsdal ML, Ryan GL. HPV and cervical dysplasia in adolescents: a progressive march toward prevention. *J Pediatr Adolesc Gynecol*. 2015;28(3):127–31.
68. Clinton SK, Giovannucci EL, Hursting SD. The World Cancer Research Fund/American Institute for Cancer Research third expert report on diet, nutrition, physical activity, and cancer: impact and future directions. *J Nutr*. 2020 Apr 1;150(4):663–71.
69. Kirsh VA, Hayes RB, Mayne ST, Chatterjee N, Subar AF, Dixon LB, et al. Supplemental and dietary vitamin E, β-carotene, and vitamin C intakes and prostate cancer risk. *J Natl Cancer Inst*. 2006;98(4):245–54.
70. Lee IM, Gaziano JM, Buring JE. Vitamin E in the prevention of prostate cancer: where are we today? *J Natl Cancer Inst*. 2006;98(4):225–227.
71. Czy U, Paradowska P. Skutki Niedoboru i Ryzyko Suplementacji Folianów w Diecie. 2010;79(6):457–63.
72. Czeczot H. Kwasfoliowy w fizjologiiipatologii folic acid in physiology and pathology. *Postepy Hig Med Dosw*. 2008;62:405–19.
73. Preventing cancer, cardiovascular disease, and diabetes | Diabetes care | American Diabetes Association [Internet]. [cited 2023 Jan 14]. Available from: https://diabetesjournals.org/care/article/27/7/1812/24572/Preventing-Cancer-Cardiovascular-Disease-and
74. Klein E, Thompson I, Tangen C, Crowley J, Scott L, Goodman P, et al. Vitamin E and the risk of prostate cancer the selenium and vitamin E cancer prevention trial (SELECT). *JAMA*. 2011;306(14):1549–56.
75. Sesso HD, Buring JE, Christen WG, Kurth T, Belanger C, MacFadyen J, et al. Vitamins E and C in the prevention of cardiovascular disease in men, the physicians' health study II randomized controlled trial. *JAMA*. 2008;300(18):2123–33.
76. Greenwald P, Anderson D, Nelson SA, Taylor PR. Clinical trials of vitamin and mineral supplements for cancer prevention. *Am J Clin Nutr*. 2007;85(1):314S–7S.
77. EPIC – European prospective investigation into cancer and nutrition [Internet]. [cited 2023 Jan 14]. Available from: https://epic.iarc.fr/
78. Kruk J. Fruits and vegetables consumption and the risk of breast cancer. *Contemp Oncol Współczesna Onkol*. 2006 Jan 1;10:224–30.
79. Atkin W. Options for screening for colorectal cancer. *Scand J Gastroenterol*. 2003 Jan 1;38(237):13–6.
80. Döbrőssy L, Kovács A, Cornides A, Budai A. Factors influencing the participation in colorectal screening. *Orv Hetil*. 2014 Jul 6;155(27):1051–6.
81. Stracci F, Zorzi M, Grazzini G. Colorectal cancer screening: tests, strategies, and perspectives. *Front Public Health*. 2014;2:210.
82. Tao Z, Shi A, Lu C, Song T, Zhang Z, Zhao J. Breast cancer: epidemiology and etiology. *Cell Biochem Biophys*. 2015;72(2):333–8.
83. Jansen LA, Backstein RM, Brown MH. Breast size and breast cancer: a systematic review. *J Plast Reconstr Aesthet Surg*. 2014;67(12):1615–23.

84. Kruk J. Overweight, obesity, oxidative stress and the risk of breast cancer. *Asian Pac J Cancer Prev.* 2014;15(22):9579–86.
85. Xia X, Chen W, Li J, Chen X, Rui R, Liu C, et al. Body mass index and risk of breast cancer: a nonlinear dose-response meta-analysis of prospective studies. *Sci Rep.* 2014;4(1):1–5.
86. Larsen MJ, Thomassen M, Gerdes AM, Kruse TA. Hereditary breast cancer: clinical, pathological and molecular characteristics. *Breast Cancer Basic Clin Res.* 2014;8:BCBCR–S18715.
87. Donepudi MS, Kondapalli K, Amos SJ, Venkanteshan P. Breast cancer statistics and markers. *J Cancer Res Ther.* 2014;10(3):506.
88. Vance RB. Common interests and common goals: achieving greater progress in preventive health through strategic collaborations. *CA Cancer J Clin.* 2004;54(4):188–9.
89. Irwin ML, Smith AW, McTiernan A, Ballard-Barbash R, Cronin K, Gilliland FD, et al. Influence of pre- and postdiagnosis physical activity on mortality in breast cancer survivors: the health, eating, activity, and lifestyle study. *J Clin Oncol.* 2008 Aug 20;26(24):3958–64.
90. Emaus A, Thune I. Physical activity and lung cancer prevention. *Phys Act Cancer.* 2010;101–33.
91. Hojman P, Gehl J, Christensen JF, Pedersen BK. Molecular mechanisms linking exercise to cancer prevention and treatment. *Cell Metab.* 2018;27(1):10–21.
92. Grossmann ME, Nkhata KJ, Mizuno NK, Ray A, Cleary MP. Effects of adiponectin on breast cancer cell growth and signaling. *Br J Cancer.* 2008;98(2):370–9.
93. Neilson HK, Friedenreich CM, Brockton NT, Millikan RC. Physical activity and postmenopausal breast cancer: proposed biologic mechanisms and areas for future research. *Cancer Epidemiol Biomark Prev Publ Am Assoc Cancer Res Cosponsored Am Soc Prev Oncol.* 2009 Jan;18(1):11–27.
94. Friedenreich CM, Woolcott CG, McTiernan A, Ballard-Barbash R, Brant RF, Stanczyk FZ, et al. Alberta physical activity and breast cancer prevention trial: sex hormone changes in a year-long exercise intervention among postmenopausal women. *J Clin Oncol Off J Am Soc Clin Oncol.* 2010 Mar 20;28(9):1458–66.
95. Leon ME, Peruga A, McNeill A, Kralikova E, Guha N, Minozzi S, et al. European code against cancer, 4th edition: tobacco and cancer. *Cancer Epidemiol.* 2015 Dec;39(Suppl1):S20–33.
96. Chapman SLC, Wu LT. E-cigarette prevalence and correlates of use among adolescents versus adults: a review and comparison. *J Psychiatr Res.* 2014;54:43–54.
97. World Health Organization, IARC. Smokeless tobacco and some tobacco-specific N-nitrosamines; 2007.
98. Diethelm PA, Farley TM. Refuting tobacco-industry funded research: empirical data shows decline in smoking prevalence following introduction of plain packaging in Australia. *Tob Prev Cessat.* 2015 Nov 27;1(1–7).
99. Becher H, Belau M, Winkler V, Aigner A. Estimating lung cancer mortality attributable to second hand smoke exposure in Germany. *Int J Public Health.* 2018;63(3):367–75.
100. Thomas JL, Guo H, Carmella SG, Balbo S, Han S, Davis A, et al. Metabolites of a tobacco-specific lung carcinogen in children exposed to secondhand or thirdhand tobacco smoke in their homes secondhand smoke exposure in children. *Cancer Epidemiol Biomarkers Prev.* 2011;20(6):1213–21.
101. Cogliano VJ, Baan R, Straif K, Grosse Y, Lauby-Secretan B, El Ghissassi F, et al. Preventable exposures associated with human cancers. *J Natl Cancer Inst.* 2011;103(24):1827–39.
102. Stefanowicz A, Kulik T, Pacian A, Dyduch A, Żo\lnierczuk-Kieliszek D. Importance of training family physicians in oncology in the prophylaxis of malignant cancer in phc. *Med Ogólna Nauki O Zdrowiu.* 2008;13(4): 294–302.

103. Dutta S, Mahalanobish S, Saha S, Ghosh S, Sil PC. Natural products: an upcoming therapeutic approach to cancer. *Food Chem Toxicol.* 2019 Jun;128:240–55.
104. Degner SC, Papoutsis AJ, Selmin O, Romagnolo DF. Targeting of aryl hydrocarbon receptor-mediated activation of cyclooxygenase-2 expression by the indole-3-carbinol metabolite 3, 3′-diindolylmethane in breast cancer cells. *J Nutr.* 2009;139(1):26–32.
105. Rahman KW, Li Y, Wang Z, Sarkar SH, Sarkar FH. Gene expression profiling revealed survivin as a target of 3, 3′-diindolylmethane-induced cell growth inhibition and apoptosis in breast cancer cells. *Cancer Res.* 2006;66(9):4952–60.
106. Riby JE, Firestone GL, Bjeldanes LF. 3, 3′-Diindolylmethane reduces levels of HIF-1α and HIF-1 activity in hypoxic cultured human cancer cells. *Biochem Pharmacol.* 2008;75(9):1858–67.
107. Wang W, Lv M, Wang Y, Zhang J. Development of novel application of 3, 3′-diindolylmethane: sensitizing multidrug resistance human breast cancer cells to γ-irradiation. *Pharm Biol.* 2016;54(12):3164–8.
108. Ahmad A, Ali S, Ahmed A, Ali AS, Raz A, Sakr WA, et al. 3,3′-Diindolylmethane enhances the effectiveness of herceptin against HER-2/neu-expressing breast cancer cells. *PLoS One.* 2013;8(1):e54657.
109. Wang Y, Gho WM, Chan FL, Chen S, Leung LK. The red clover (Trifolium pratense) isoflavone biochanin A inhibits aromatase activity and expression. *Br J Nutr.* 2008;99(2):303–10.
110. Sehdev V, Lai JC, Bhushan A. Biochanin A modulates cell viability, invasion, and growth promoting signaling pathways in HER-2-positive breast cancer cells. *J Oncol.* 2009;2009. Article ID 121458.
111. Moon YJ, Shin BS, An G, Morris ME. Biochanin A inhibits breast cancer tumor growth in a murine xenograft model. *Pharm Res.* 2008;25(9):2158–63.
112. Liu Q, Loo WT, Sze SCW, Tong Y. Curcumin inhibits cell proliferation of MDA-MB-231 and BT-483 breast cancer cells mediated by down-regulation of NFκB, cyclinD and MMP-1 transcription. *Phytomedicine.* 2009;16(10):916–22.
113. Kakarala M, Brenner DE, Korkaya H, Cheng C, Tazi K, Ginestier C, et al. Targeting breast stem cells with the cancer preventive compounds curcumin and piperine. *Breast Cancer Res Treat.* 2010;122(3):777–85.
114. Chen Y, Shu W, Chen W, Wu Q, Liu H, Cui G. Curcumin, both histone deacetylase and p300/CBP-specific inhibitor, represses the activity of nuclear factor kappa B and Notch 1 in Raji cells. *Basic Clin Pharmacol Toxicol.* 2007 Dec;101(6):427–33.
115. Yang J, Cao Y, Sun J, Zhang Y. Curcumin reduces the expression of Bcl-2 by upregulating miR-15a and miR-16 in MCF-7 cells. *Med Oncol.* 2010 Dec;27(4):1114–8.
116. Aggarwal BB, Shishodia S, Takada Y, Banerjee S, Newman RA, Bueso-Ramos CE, et al. Curcumin suppresses the paclitaxel-induced nuclear factor-κB pathway in breast cancer cells and inhibits lung metastasis of human breast cancer in nude mice. *Clin Cancer Res.* 2005 Oct 20;11(20):7490–8.
117. Holzer TR, McMaster WR, Forney JD. Expression profiling by whole-genome interspecies microarray hybridization reveals differential gene expression in procyclic promastigotes, lesion-derived amastigotes, and axenic amastigotes in Leishmania mexicana. *Mol Biochem Parasitol.* 2006 Apr;146(2):198–218.
118. Takeshima M, Ono M, Higuchi T, Chen C, Hara T, Nakano S. Antiproliferative and apoptosis-inducing activity of lycopene against three subtypes of human breast cancer cell lines. *Cancer Sci.* 2014;105(3):252–7.
119. Peng SJ, Li J, Zhou Y, Tuo M, Qin XX, Yu Q, et al. In vitro effects and mechanisms of lycopene in MCF-7 human breast cancer cells. *Genet Mol Res.* 2017 Apr 13;16(2).
120. Verhoeven DT, Goldbohm RA, van Poppel G, Verhagen H, van den Brandt PA. Epidemiological studies on brassica vegetables and cancer risk. *Cancer Epidemiol Biomark Prev Publ Am Assoc Cancer Res Cosponsored Am Soc Prev Oncol.* 1996;5(9):733–48.

121. Weng JR, Tsai CH, Kulp SK, Chen CS. Indole-3-carbinol as a chemopreventive and anti-cancer agent. *Cancer Lett.* 2008;262(2):153–63.
122. Fan S, Meng Q, Saha T, Sarkar FH, Rosen EM. Low concentrations of diindolylmethane, a metabolite of indole-3-carbinol, protect against oxidative stress in a BRCA1-dependent manner. *Cancer Res.* 2009;69(15):6083–91.
123. Licznerska BE, Szaefer H, Murias M, Bartoszek A, Baer-Dubowska W. Modulation of CYP19 expression by cabbage juices and their active components: indole-3-carbinol and 3, 3′-diindolylmethene in human breast epithelial cell lines. *Eur J Nutr.* 2013;52(5):1483–92.
124. Marques M, Laflamme L, Benassou I, Cissokho C, Guillemette B, Gaudreau L. Low levels of 3, 3′-diindolylmethane activate estrogen receptor α and induce proliferation of breast cancer cells in the absence of estradiol. *BMC Cancer.* 2014;14(1):1–10.
125. Heinonen SM, Wähälä K, Adlercreutz H. Identification of urinary metabolites of the red clover isoflavones formononetin and biochanin A in human subjects. *J Agric Food Chem.* 2004;52(22):6802–9.
126. Atwell LL, Zhang Z, Mori M, Farris PE, Vetto JT, Naik AM, et al. Sulforaphane bioavailability and chemopreventive activity in women scheduled for breast biopsy. *Cancer Prev Res (Phila PA).* 2015;8(12):1184–91.
127. Moon YJ, Brazeau DA, Morris ME. Effects of flavonoids genistein and biochanin A on gene expression and their metabolism in human mammary cells. *Nutr Cancer.* 2007;57(1):48–58.
128. Wang Y, Yu J, Cui R, Lin J, Ding X. Curcumin in treating breast cancer: a review. *J Lab Autom.* 2016;21(6):723–31.
129. Choudhuri T, Pal S, Agwarwal ML, Das T, Sa G. Curcumin induces apoptosis in human breast cancer cells through p53-dependent Bax induction. *FEBS Lett.* 2002;512(1–3):334–40.
130. Zong H, Wang F, Fan Q xia, Wang L xing. Curcumin inhibits metastatic progression of breast cancer cell through suppression of urokinase-type plasminogen activator by NF-kappa B signaling pathways. *Mol Biol Rep.* 2012;39(4):4803–8.
131. Lindvall C, Bu W, Williams BO, Li Y. Wnt signaling, stem cells, and the cellular origin of breast cancer. *Stem Cell Rev.* 2007;3(2):157–68.
132. Liu S, Dontu G, Wicha MS. Mammary stem cells, self-renewal pathways, and carcinogenesis. *Breast Cancer Res.* 2005 Mar 30;7(3):86.
133. Chalabi N, Delort L, Le Corre L, Satih S, Bignon YJ, Bernard-Gallon D. Gene signature of breast cancer cell lines treated with lycopene. *Pharmacogenomics.* 2006 Jul;7(5):663–72.
134. King-Batoon A, Leszczynska JM, Klein CB. Modulation of gene methylation by genistein or lycopene in breast cancer cells. *Environ Mol Mutagen.* 2008;49(1):36–45.
135. Bishop KS, Ferguson LR. The interaction between epigenetics, nutrition and the development of cancer. *Nutrients.* 2015 Feb;7(2):922–47.
136. Eliassen AH, Hendrickson SJ, Brinton LA, Buring JE, Campos H, Dai Q, et al. Circulating carotenoids and risk of breast cancer: pooled analysis of eight prospective studies. *J Natl Cancer Inst.* 2012 Dec 19;104(24):1905–16.
137. Mitra S, Dash R. Natural products for the management and prevention of breast cancer. Evidence-Based Complementary and Alternative Medicine. 2018;2018(1):8324696.

6 *Amaranthus* and Sea Buckthorn
Future Smart Crops for Cancer Therapy

Yengkhom Linthoingambi Devi, Biseshwori Thongam, and Rajkumari Jashmi Devi
Institute of Bioresources and Sustainable Development (IBSD), Imphal West, India

6.1 INTRODUCTION

Amaranthus and sea buckthorn are one of the potential 'neglected and underutilized species (NUS)'. Neglected and underutilized species (NUS) are species with under-exploited potential for contributing to food security, health (nutritional/medicinal), income generation and environmental services (Jaenicke and Hoeschle-Zeledon, 2006). They are also referred as orphan or minor or underutilized or traditional species to characterize the range of plant species. NUS are considered important in local consumption and production systems. They are grown by farmers traditionally in their centres of diversity which helped in supporting the nutritional security and livelihood goals. They have high adaptation to agro-ecological niches and marginal areas; however, they are ignored by policy-makers and excluded from R&D agendas which result in the loss of their diversity and associated traditional knowledge with it. They are commonly represented by wild species, ecotypes and landraces; they are cultivated and utilized by drawing on indigenous knowledge (Padulosi et al., 2007). NUS crops are well known for their adaptation to multiple stress conditions (Massawe et al., 2016). While NUS may be well adapted to multiple stress conditions, they are grown in geographical pockets that are often far from where they could provide the most positive contribution to food and nutrition security.

The genus *Amaranthus* of Amaranthaceae family is a native crop of tropical America. The crop is used as a staple crop in Mayan, Aztec and Incan civilizations. It is widely consumed and cultivated in Nepal, India, Indonesia, China, Philippines, Malaysia, Mexico, Southern and Central America. Some of the common names of *Amaranthus* are pigweed, purple amaranth, red amaranth, kiwicha and alegria. *Amaranthus* is a multipurpose crop with a resistance to abiotic factors like heat stress and drought condition and also to a number of pest conditions with a short production cycle. The adaptability of this crop to a new area is much higher than other

DOI: 10.1201/9781003477792-6

cereal crops. Because of which it is known as golden vegetables with high economic importance (Hoidal et al., 2020). The genus *Amaranthus* is represented by around 70 species. Of which, 17 species are edible and three are grain amaranths. While *Amaranthus tricolor, Amaranthus blitum, Amaranthus spinosus* and *Amaranthus viridis* are known for their edible leaves; *Amaranthus caudatus, Amaranthus cruentus* and *Amaranthus hypochondriacus* are known for their grains. *A. viridis* is the most cultivated species for consumption. It is also the most commonly found species in India (Ebert et al., 2011). Among the species, *A. cruentus* is known for leaves and grains. According to Das (2012), some of the *Amaranthus* species are categorized as weedy species. They are *A. spinosus, A. viridis, Amaranthus retroflexus, A. graecizans, Amaranthus dubius* and *Amaranthus hybridus.*

The flour of *Amaranthus* are used as whole and also mixed with other cereals like sorghum, maize and millet. It is used for the production of several confectionary items like pancakes, crackers, paste, muffins and breakfast cereals. The grains are consumed as porridge and also could be malted for beer production (Emire and Arega, 2012). Amaranthus possessed several properties like anti-inflammatory, anti-diabetic, anti-malarial, anti-depressant, anti-pyretic, anti-anaphylactic, anti-hypercholestrolemic, anti-carcinogenic, hepatoprotective, anti-inflammatory, anti-bacterial, anti-microbial and anti-tumour.

Sea buckthorn of the Elaeagnaceae family has a large number of species of about 100, of which 12 subspecies are acknowledged for cultivation. Some of the commonly cultivated subspecies of sea buckthorn for food and medicinal properties are sea buckthorn subsp. *carpatica*, sea buckthorn subsp. *caucasia*, sea buckthorn subsp. *sinensis*, sea buckthorn subsp. *mongolia*, sea buckthorn subsp. *rhamnoides*, sea buckthorn subsp. *sinensis*, sea buckthorn subsp. *Turkestanica* and sea buckthorn subsp. *yunnanensis.* The crop could survive under a wide range of temperatures ranging from −43°C to 40°C and a wide range of soil conditions. The crop has been used in different countries as a source of food and a source of medicine (Li and Beveridge, 2003; Ruan et al., 2013). According to ancient Chinese, the crop possesses the power to remove phlegm, nourishes stomach, benefits lung, promotes blood circulation, strengthens spleen and removes blood stasis (Li et al., 2019). Sea buckthorn has several properties like cardioprotective, anti-oxidant, anti-diabetic, anti-atherogenic, hepatoprotective, anti-cancerous, immunomodulating, anti-bacterial, anti-virus, vasodilating effect and anti-inflammatory.

The fruit pulp and berries have been used in the preparation of jams, squash, juices, marmalades, pies and liquors (Hu, 2005). Oil can be extracted from the seed of sea buckthorn. Teas can be prepared from leaves, seeds and fruits. The teas made from seed helped in weight loss and have laxative property, while teas made from leaves have anti-diarrheal property and teas made from fruits are effective against several skin diseases (Frohne, 2010).

6.1.1 Nutritional Components of Different *Amaranthus* Species and Sea Buckthorn

Amaranthus is a highly nutritional gluten-free pseudocereal (de la Hey and Beinart, 2017). The crop has been used from ancient times due to its excellent nutritional content in leaves and grains. Protein content in leaves and grains of amaranth are reported to be about 14%–30% and 15%–43%, respectively, of fresh matter. The amino acid content of amaranth is well-balanced with higher content of unsaturated fatty acids (The World Bank, Statistics SA, 2018). An essential amino acid, lysine, is present in *Amaranthus* which is absent in many other grains (Amare et al., 2015). The protein content of *Amaranthus* has been reported near to the levels recommended by the Food and Agricultural Organization (FAO) and World Health Organization (WHO) due to its balanced content of amino acids (O'Brien and Price, 1983; Murya and Pratibha, 2018). Important sulphur-containing amino acids like methionine and cysteine are also contained in higher amount in proteins of *A. cruentus* (Escudero et al., 2004; Schmidt and Fedtke, 1977; Martínez-Núñez et al., 2019).

Amaranthus is also considered as a storehouse of vitamins, precursors of vitamins (riboflavin, ascorbic acid, carotenoids and tocols) and minerals (K, Fe, Ca, Mg, Cu, Mn and Zn) which is much higher than other leafy vegetables and cereals (Musa et al., 2011). These rich nutrients composition makes the plant an excellent ingredient of food for producing gluten-free products (Mabhaudhi et al., 2019; Baldermann et al., 2016). *Amaranthus* is also a rich source of phytochemicals which include flavonoids, alkaloids, saponins, glycosides, steroids, phenolic acids, vitamins, amino acids, minerals, terpenoids, lipids, betaine, catechuic tannins and carotenoids (Nana et al., 2012). Dlamini et al. (2010) reported *A. cruentus* as a potential source of pro-vitamin A carotenoid (β-carotene) with highest content in leaves followed by seeds, stem and roots. The carotenoid content in *A. cruentus* was reported to be seven times higher than that in tomatoes. Canthaxanthin contributed the major carotenoid followed by β-carotene and lutein. *Amaranthus* species differ in the amount of nutritional components content present in them. For example, the calcium content in the dry leaf sample of *A. spinosus* was found higher (4,500/100 g dry weight) than other species like *A. tricolor, A. viridis* and *A. blitum*, while *A. viridis* had highest iron content followed by *A. spinosus, A. tricolor* and *A. blitum.*

Berries and other part of sea buckthorn are a rich source of nutrients and biologically active compounds. The fruits are oval shaped and 6–9 mm long and are found in orange, dark yellow or red colour. Fruit pulp is present between the fruit and the seed which is a juicy, soft and fleshy tissue (Michel et al., 2012). The berries of sea buckthorn are a rich source of minerals like Ca, K, Fe and P. It also contains a good source of vitamin C in the range from 360 to 2,500 mg/100 g (Bal et al., 2011). Besides this, the plant is also a good source of vitamin E (Michel et al., 2012), vitamin B like thiamine and riboflavin (Christaki, 2012) and vitamins A and K (Fatima et al., 2012). Different carotenoids are also available in the berries of the fruit like lycopene; α-, β- and γ-carotenes; glycopene; zeaxanthin; β-carotene and lutein (Michel et al., 2012). Several organic acids like malic acid, oxalic acid, quinic acid, citric acid and tartaric acid are also present in sea buckthorn. Flavonoids like kaempferol, quercetin, myricetin and isorhamnetin are also present in rich amount.

The peel of fruits and stem contain 5-hydroxytryptamine, which is used for post-shock depression treatment (Kumar et al., 2011). Phenolic compounds are the main content in the leaves of sea buckthorn. The leaves also contain polyphenols, catechin, carotenoid lycopene, catechin, bioflavonoids and coumarins. Oil can also be extracted from sea buckthorn either from seeds or from the pulp. The pulp contains about 4%–13% oil with rich contents of carotenoids, carotene, vitamin E and unsaturated fatty acids, mainly linolenic acid and linoleic acid (Christaki, 2012). While the oil extracted from the seed contains about 8%–20%, with linoleic acid, linolenic acid and oleic acid as main unsaturated fatty acids contributing 90% and the remaining 10% by saturated fatty acids like palmitic acid (Valíček and Havelka, 2008).

6.1.2 Anti-Nutritional Component of *Amaranthus* and Sea Buckthorn

Despite having a good nutritional content, *Amaranthus* also present some anti-nutritional factors. Some of them are phytate, tannins, saponins, protease inhibitors, oxalates and nitrates. Dobos (1992) found anti-nutrient factors such as saponin and protease inhibitors such as trypsin and chymotrypsin in low concentration. Oxalates and phytates affect nutrient inhibition in amaranth grain. A high amount of tannins is present mostly in grain hull; so, the removal of grain hull gets rid of its presence. Phytic acid on the other hand is found distributed uniformly in the grain and its removal is done by abrasive method and water extraction. These anti-nutritional factors, even though they are not good for consumption, help in the protection of the plant through different processes. Phytic acid, an anti-nutritional factor, when consumed by humans, cannot be digested properly by our body, but they provide a source of storage of phosphorus in plant cell. Also, saponins can reduce the absorption of mineral elements like iron and zinc by binding to them (Cuadrado et al., 2019). These anti-nutrients can, however, be removed through different methods like defatting, germination, blanching, autoclaving and fermentation to improve their digestibility and also to reduce anti-nutrient factors (Olawoye and Gbadamosi, 2017).

Sea buckthorn contains tannins (8%) in the leaves (Valíček and Havelka, 2008). Tannins have a molecular weight of more than 500 and are water-soluble phenolic compounds which have the ability to precipitate proteins from aqueous solution. The anti-nutritional property of tannins in the diet is because of their ability to create interactions with dietary proteins. So, the tannin-protein complex is responsible for low digestibility, growth depression and increased faecal nitrogen content. Besides this, tannins also inhibit enzymes like trypsin and amylases which are involved in digestion process (Davis, 1982).

6.1.3 Bioactive Components and Secondary Metabolites of *Amaranthus* and Sea Buckthorn

Besides having a good amount of protein, vitamins, minerals and phytochemicals, *Amaranthus* also contains secondary metabolites like several bioactive components, phenolic compounds which are beneficial to human diet. The phenolic compounds like benzoic acid, hydroxycinnamic acid, flavonols and their glycosides are found in leaves, flowers and stalk of *Amaranthus* (Neugart et al., 2017; Li et al., 2015). Leaves of *A. hypochondriacus* and *A. tricolor* contained a high amount of betaxanthins and betacyanins. The most abundant flavonoids present in them are quercetin and rutin while common phenolic acid present are syringic, salicylic, gallic, vanilic, ferulic, p-coumaric, sinapic and ellagic acid (Khanam and Oba, 2013). Methyl derivatives of arginine betaxanthin and betalamic acid were present in the leaves of *A. tricolor.*

The leaves at the maturity stage of *Amaranthus* species are a good source of quercetin and its derivatives. Rutin is present in high amount in *A. viridis* and *A. cruentus* (Kalinova and Dadakova, 2009). *A. spinosus* contains phytochemicals mostly in the root of the plant in the form of alkaloids, terpenes, glycosides and sugars as reported by Jhade et al. (2011). Bioactive compounds are known for its health-promoting effect and are found in amaranth grain (Karamać et al., 2019). Main bioactive compounds present in *Amaranth* grain are rutin, phenolic acids like hydroxybenzoic, protocatechuic, ferulic and caffeic acid, isoquercitin and nicotiflorin. Bioactive compounds like nicotiflorin protect memory function; the prevention of oxidation by quercetin and rutin helps in the reduction of ageing process (Alvarez-Jubete et al., 2010). Phytopigments and phytochemicals like betaxanthins, betacyanins, chlorophylls, carotenoids, β-carotene, vitamin C, phenolics and flavonoids are some of the naturally available anti-oxidants in amaranth vegetable (Venskutonis and Kraujalis, 2013; Repo-Carrasco-Valencia et al., 2010). Regular consumption of *Amaranthus* is believed to reduce the blood cholesterol level and also help people with cardiovascular disease and hypertension (Olaniyi, 2007; Kolawole and Sarah, 2009). *Amaranthus caudatus* has anti-cancerous property with cholesterol reducing effect and improves liver function (Chávez-Jáuregui et al., 2000). Aerial extracts of *A. hybridus* and *A cruentus* are a good source of xanthine oxidase inhibitor and anti-oxidants. Polyphenols are widely present in consumed foods which are bioactive secondary metabolites and anti-oxidants, and its three main types are flavonoids, tannins and phenolic acids. These compounds are a boon to several diseases like diabetes and osteoporosis. Betalains is another secondary metabolite and bioactive compound found mostly in plants like *Amaranthus*, cactus and beetroot. Condensation product of betalamic acid with various other amino acid compounds gives betaxanthins which is the chemical constituent of betalains (Stintzing et al., 2004). Cai et al. (2003) also found that plants of the Amaranthaceae family are packed with anti-oxidants like betaxanthins and betacyanins. Additional hydroxycinnamyl was found in the aerial parts of young plants (Steffensen et al., 2011). Leaf extracts are a good source of phenolics and scavenged free radicals and also inhibit the production of nitric acid as found by Conforti et al. (2005).

The major bioactive components present in sea buckthorn are phenolic acids, flavonoids, proanthocyanidins, fatty acids, carotenoids, triterpenoids, vitamins (vitamins C, K and E), lignans and phytosterols. More than 60 flavonoids and 10 phenolic acids were present in sea buckthorn. Flavonols and isorhamnetin derivatives are the dominant flavonoids present, contributing about 98% of flavonoids which is followed by quercetin derivatives (Tkacz et al., 2020). Isorhamnetin has anti-inflammatory activity. Ursolic acid, oleanolic acid and casuarinin are also important anti-oxidant and anti-inflammatory agents present in sea buckthorn (Rédei et al., 2018; Qi et al., 2018). The methanol extract (Liu et al., 2015) and phenolic-rich fraction of crude extract of sea buckthorn leaves have higher anti-bacterial property than chloroform and acetone extracts against *Shigella dysenteriae* (Kumar et al., 2013). Flavonoids and unsaturated fatty acids contained in seed berry oil improved the cardiovascular system (Suryakumar and Gupta, 2011) and protected from heart diseases (Christaki, 2012). The phenolic compounds of sea buckthorn exhibit anti-bacterial and anti-viral effects which reduce gram-positive bacteria (Kumar and Sagar, 2007) and suppress gram-negative bacteria (Khan et al., 2010). Michel et al. (2012) reported that the active agent in sea buckthorn inhibits certain bacteria like *Pseudomonas aeruginosa*, *Staphylococcus aureus*, *Bacillus cereus* and *Yersinia enterocolitica*. Sea buckthorn contains hepatotoxins like carbon tetrachloride, ethanol and acetaminophen which cause degeneration, damage and subsequent death of hepatic cell (Solcan et al., 2013). Also, bioactive components like α-tocopherol, β-carotene and unsaturated fatty acids also protect hepatic cells from hepatotoxins (Ramesbabu et al., 2011). Maheshwari et al. (2011) studied the leaves of sea buckthorn to understand the phenolic constituents through reverse-phase high-performance liquid chromatography (RP-HPLC) and found phenolic compounds like myricetin, gallic acid, quercetin, isorhamnetin and kaempferol. Chemical components of sea buckthorn like quercetin, isorhamnetin and quassin strengthen the immune system of organism and thus elevate the resistance to illness (Li and Beveridge, 2003). The products of sea buckthorn like hexane also actively participate against stress and indomethacin (Khan et al., 2010).

6.2 ANTI-CANCEROUS PROPERTIES OF *AMARANTHUS* AND SEA BUCKTHORN

Secondary metabolites and bioactive compounds like quercetin, kaempferol, p-coumaric acid, syringic acid and ferulic acid are some important anti-cancerous compounds present in *Amaranthus*. *A. caudatus* has a high amount of flavonoid composition of kaempferol and quercetin which is known for their anti-oxidant and anti-inflammatory property in addition to anti-cancerous property. Kaempferol helps in lowering susceptibility of humans to oxidative stress and reactive oxidative species (ROS) level, while quercetin reduces oxidative damage to biomolecules and lowers lipid peroxidation (Askari et al. 2013; Leonov et al. 2015). Quercetin and rutin are aglycone form of number of flavonoid glycosides and have properties to prevent cancer, heart diseases and blood vessel diseases (Holland et al., 2020). Through quenching of free radicals in the body, quercetin protects the oxidation of low-density lipoprotein (Cartea et al., 2010). Quercetin is reported to inhibit the production of LPS-induced IL-8 in lung A549 cells and the production of lipopolysaccharide (LPS)-induced tumour necrosis factor α (TNF-α) in macrophages (Khushnuma et al., 2021). Kaempferol, a natural flavonol present in *Amaranthus,* is another anti-cancerous metabolite. Kaempferol (KMF) is reported to inhibit the proliferation of various cancer cells by arresting the transition of one cell cycle to another (Dubey et al., 2021; Meher et al., 2022; Patnaik et al., 2022). Another anti-cancerous compound in *Amaranthus* is p-coumaric acid (p-CA). p-CA is reported to limit the proliferation of tumour cells. It encouraged the death of tumour cells and triggered apoptosis in melanoma cells to control their proliferation (Cláudia et al., 2015). Syringic acid is another natural origin phenolic substance found in *Amaranthus*. Syringic acid is used for the treatment of a number of diseases like cardiovascular disease, diabetes, cancer, neuropathy, cerebral ischaemia and liver damage (Croft, 1998). Ferulic acid extracted from *A. viridis* is used against lung cancer and tumours of the central nervous system (Popoola, 2022). Ellagic acid also has anti-carcinogenic and anti-mutagenic responses (Maas et al., 1991).

Ethyl-acetate, hexane and methanolic extracts of *A. tristis* are reported as anti-cancerous as they showed anti-proliferative properties with fewer side effects in human colon adenocarcinoma cell line (COLO-320-DM) (Baskar et al., 2012). Ethanolic extracts of leaves of *A. spinosus* is used against cancer (Joshua et al., 2010). Bulbul et al. (2011) studied the leaf extracts of *A. spinosus* and found high cytotoxic ability against diseases like cancers and tumours. Extracts of *A. caudatus* and *A. hybridus* protect detoxifying enzymes like alkaline phosphatase and gamma glutamyl transferase (GGT) as it possessed anti-carcinogenic properties. Seed protein of *Amaranthus mantegazzianus* is reported to induce necrosis and apoptosis of UMR106 cell line exhibiting potential anti-tumour properties (Barrio and Añón, 2010).

Cancer preventive peptides similar to soybean lunasin, a known anti-cancerous peptide which helps in histone acetylation of H3 by H4 and transformation of NIH-3T3 cells to cancerous foci, was found in *Amaranthus* by Maldonado et al. (2010). Determination of lectin activity of the *Amaranthus* extract was carried out using haemagglutination assay on mouse blood. Seeds extracts of *A. hybridus* and *Amaranthus lividus* were applied on 36 Swiss albino mice having Ehrlich's ascites carcinoma (EAC) cells and caused mitochondrial apoptosis of EAC cells (Al-Mamun et al., 2016). Larbie et al. (2015) reported that 50% ethanolic extracts of *A. viridis* possessed anti-proliferative property against HL-603 (human leukaemic cell lines).

Amornrit and Santiyanont (2016) induced oxidative stress using H_2O_2 in human neuroblastoma cell lines (SH-SY5Y). They also performed pretreatment of the leaf extracts of *A. tricolor* and *A. lividus* using methanol, petroleum ether and dichloromethane and a decrease in toxicity of cell was observed.This extract downregulatedproduction of intracellular ROS by downregulating several genes like RAGE, HMOX-1 and RelA/NF-jB in SH-SY5Y cells.

Phytopharmaceutical compounds which have anti-cancer properties in sea buckthorn are curcumin, resveratrol, proanthocyanidins, procyanidins and flavonoids (Shanmugam et al., 2015; Ko et al., 2017; Chen et al., 2014; Wang et al., 2014; Kristo et al., 2016). Teng et al. (2006) found a flavonoid, isorhamnetin, effective against hepatocellular carcinoma cells (BEL-7402) with a concentration of 75 μg/mL after 72 hours of treatment. Li et al. (2015) studied the lung cells of mice, i.e., C57BL/6 for 7 days using different concentrations of isorhamnetin ranging from 10 to 320 μg/mL and found to have anti-proliferation effect. Wang et al. (2014) used procyanidins isolated from the seeds of sea buckthorn on human breast cancer cell, MDA-MB-231 and found inhibitory effect on fatty acid synthase (FAS) which is found in cancer cells. A concentration of 0.087 μg/mL was found effective for 50% FAS activity. It was further report that procyanidin can induce apoptosis of this cell by prom oting intracellular FAS activity. Guo et al. (2017) studied the anti-oxidant and anti-proliferative properties of four different species against HepG2 human liver cells in vitro and the highest activity was found in *H ippophae rhamnoides* L. subsp. *yunnanensis*. Higher amount of phenolic acid and flavonoid aglycones was also found in *H ippophae rhamnoides* L. subsp. *yunnanensis*. Zhamanbaeva et al. (2014) studied the effect of differentiation and growth of human acute leukaemia cells (KG-1a, HL60 and U937) by using ethanol extracts from the leaves of sea buckthorn and found that the extracts can inhibit cell growth based on the concentration of the extract.

They found out that the anti-proliferative activity of the extracts was obtained by the activation of S phase checkpoint and apoptosis of the cell. Kim et al. (2017) studied the phenolic and catechin contents of leaf extract of sea buckthorn in rat C6 glioma cells and found inhibition of the cell by performing apoptosis. They further report that there was a decrease in the production of reactive oxygen species because of which the proliferation process could be hindered. Oil extracted from seed of sea buckthorn was also found to retard the growth of tumour up to 3%–50% as reported by Wang et al. (1989) and also used against cervical cancer (Wu et al. 1989). Zhang (1989) found that seed oil injection inhibited the growth rate of sarcoma (S_{180}) tumour and melanoma (B_{16}) tumour in mice. Yasukawa et al. (2009) studied 70% ethanol extract from sea buckthorn branches which contains phenolic compounds like gallocatechin, catechin and epigallocatechin along with epigallocatechin and ursolic acid against two groups of mice and was found to have anti-cancer property. Besides the phenolic compounds of sea buckthorn, other compounds like high methoxyl homogalacturonana HRWP-A, a water-soluble homogenous polysaccharide, also have anti-cancer property. This compound was found to inhibit Lewis lung carcinoma (LLC) growth in mice bearing tumour.

Amaranthus species and its bioactive components having anti-cancerous property were briefly presented in Table 6.1. Anti-cancerous compounds reported from different plant parts of sea buckthorn are briefly presented in Table 6.2.

TABLE 6.1
List of Different Species of *Amaranthus* and Their Bioactive Components Having Anti-Cancerous Properties

Scientific Name	Plant Part Used	Active Component	Pharmacological Activity	References
Amaranthus cruentus	Leaves and stem	Aqueous	Anti-cancer	Gandhi et al. (2011)
	Leaves and stem	50% ethanolic extract	Anti-proliferative	Larbie et al. (2015)
Amaranthus tristis Roxb.	Root	Hexane, ethyl-acetate,	Anti-cancer	Baskar et al. (2012)
Amaranthus mantegazzianus	Seeds	Protein isolate	Anti-tumour	Barrio and Añón (2010)
Amaranth (*Amaranthus cruentus*)	Seeds	Isolated peptide	Anti-cancer	Maldonado-Cervantes et al. (2010)
Amaranthus caudatus	-	Kaempferol and quercetin	Anti-cancer	Miraj (2016)
Amaranthus viridis	Leaves	Ethanolic extracts	Anti-proliferative property	Larbie et al. (2015)

TABLE 6.2
Anti-Cancer Compounds of Sea Buckthorn

Plant Part Used	Anti-Cancerous Compound	Cancer Cells	Effect	References
Berries	Ethanol water extract	Cancer cells of the breast and semi-colon	Inhibition of tumour cell proliferation of all tested lines	Olsson et al. (2004)
Berries	Ethanol extract	Acute myeloid leukaemia cells	Anti-proliferative action	Zhamanbaeva et al. (2014)
Berries	Isorhamnetin	Cancer cells of the liver	Cytotoxicity against cancer cells	Teng et al. (2006)
Berries	Isorhamnetin	Lung cancer cells	Anti-proliferative action	Li et al. (2015)
Seeds	Procyanidins	Breast cancer cells	Inducing apoptosis	Wang et al. (2014)
Leaf extracts	Phenolic acid and flavonoid aglycones	HepG2 human liver cells	Anti-oxidant and anti-proliferative properties	Guo et al. (2017)
Leaf extracts	Phenolic and catechin contents	C6 glioma cells in rat	Apoptosis	Kim et al. (2017)
Branches	70% ethanol extract and phenolic compounds	Mice cancer cell	Anti-cancer property	Yasukawa et al. (2009)

6.3 CONCLUSION

Amaranthus is a crop packed with lots of anti-cancerous active compounds. Besides its anti-cancerous activity, *Amaranthus* has different species possessing different properties which could be used against different diseases like cardiovascular diseases, atherosclerosis, arthritis, cataracts, retinopathy, neurodegenerative diseases and emphysema. The plant also possessed resistance to several diseases and pests. The plant has become an interest of study to scientists in recent years. Still many bioactive compounds are to be discovered from different species of *Amaranthus*. More exploration on different species of *Amaranthus* is still needed to study the content of bioactive compounds in them. More and more research work is conducted and needs to be conducted to understand anti-cancerous materials and the pharmacological potential present in them.

Based on the results of several experiments, sea buckthorn contains potential anti-cancerous compounds which could be used for the treatment and prevention of cancer. Besides this, it can also be used in accelerating the recovery rate of patients after receiving chemotherapy by improving their immune system. The bioactive

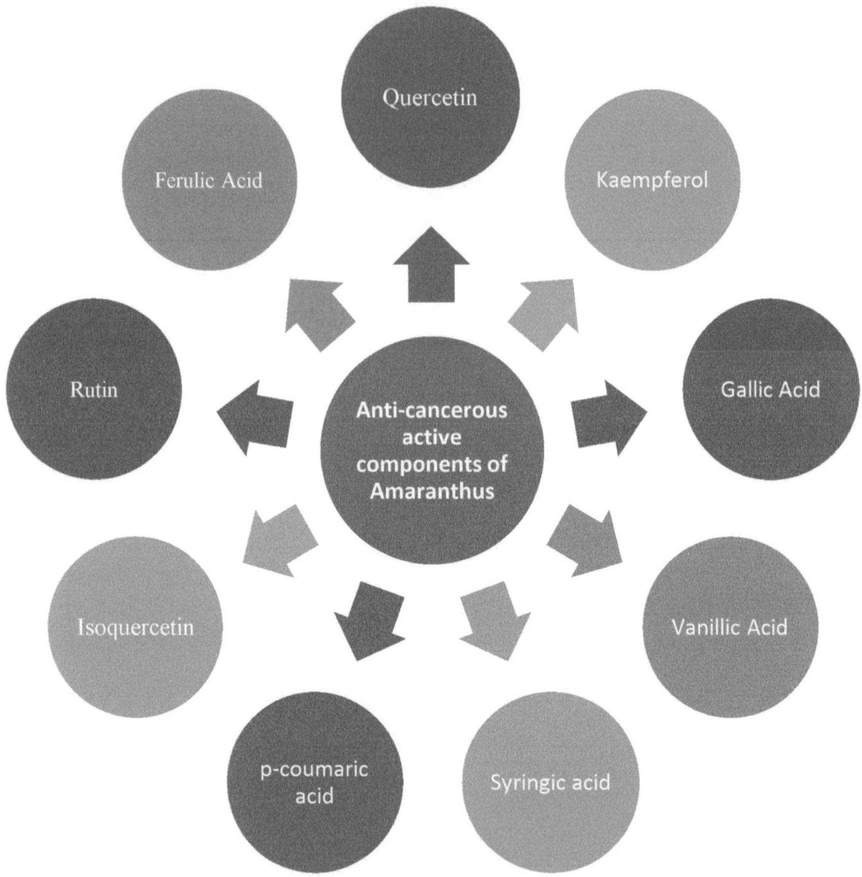

FIGURE 6.1 Diagrammatic representation of anti-cancerous active components present in different species of *Amaranthus*.

substances obtained from several parts of sea buckthorn have several properties like anti-inflammatory, anti-oxidant and anti-proliferative activities because of which the crop is known for 'golden mean' for the treatment of cancer. Further studies on prophylactic doses and more medicinal properties still need to be studied (Figures 6.1 and 6.2).

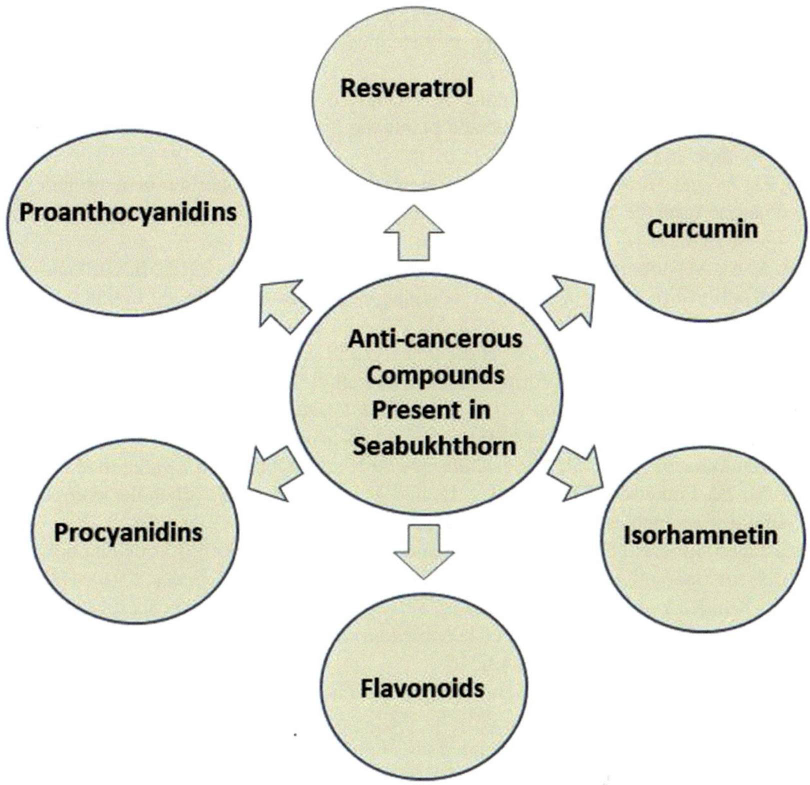

FIGURE 6.2 Diagrammatic representation of anti-cancerous active components present in different parts of sea buckthorn.

REFERENCES

Al-Mamun, M. A., Husna, J., Khatun, M., Hasan, R., Kamruzzaman, M., Hoque, K. M. F., Reza, M. A., and Ferdousi, Z. (2016). Assessment of antioxidant, anticancer and antimicrobial activity of two vegetable species of Amaranthus in Bangladesh. *BMC Complementary and Alternative Medicine*, *16*, 1–11.

Alvarez-Jubete, L., Wijngaard, H., Arendt, E. K., and Gallagher, E. (2010). Polyphenol composition and in vitro antioxidant activity of amaranth, quinoa buckwheat and wheat as affected by sprouting and baking. *Food Chemistry*, *119*(2), 770–778.

Amare, E., Mouquet-Rivier, C., Servent, A., Morel, G., Adish, A., and Haki, G. D. (2015). Protein quality of amaranth grains cultivated in Ethiopia as affected by popping and fermentation. *Food and Nutrition Sciences*, *6*(1), 38.

Amornrit, W. and Santiyanont, R. (2016). Neuroprotective effect of Amaranthus lividus and Amaranthus tricolor and their effects on gene expression of RAGE during oxidative stress in SH-SY5Y cells. *Genetics and Molecular Research*, *15*, 15027562.

Askari, G., Ghiasvand, R., Paknahad, Z., Karimian, J., Rabiee, K., Sharifirad, G., and Feizi, A. (2013). The effects of quercetin supplementation on body composition, exercise performance and muscle damage indices in athletes. *International Journal of Preventive Medicine*, *4*, 21–26.

Bal, L. M., Meda, V., Naik, S. N., and Satya, S. (2011). Sea buckthorn berries: a potential source of valuable nutrients for nutraceuticals and cosmoceuticals. *Food Research International*, *44*(7), 1718–1727.

Baldermann, S., Blagojević, L., Frede, K., Klopsch, R., Neugart, S., Neumann, A., and Schreiner, M. (2016). Are neglected plants the food for the future? *Critical Reviews in Plant Sciences*, *35*(2), 106–119.

Barrio, D. A. and Añón, M. C. (2010). Potential antitumor properties of a protein isolate obtained from the seeds of Amaranthus mantegazzianus. *European Journal of Nutrition*, *49*(2), 73–82.

Baskar, A. A., Al Numair, K. S., Alsaif, M. A., and Ignacimuthu, S. (2012). In vitro antioxidant and antiproliferative potential of medicinal plants used in traditional Indian medicine to treat cancer. *Redox Report*, *17*, 145–56.

Bulbul, I. J., Nahar, L., Ripa, F. A., and Haque, O (2011). Antibacterial, cytotoxic and antioxidant activity of chloroform, n-hexane and ethyl acetate extract of plant *Amaranthus spinosus*. *International Journal of PharmTech Research*, *3*(3), 1675–1680.

Cai, Y., Sun, M., and Corke, H. (2003). Antioxidant activity of betalains from plants of the Amaranthaceae. *Journal of Agricultural and Food Chemistry*, *51*(8), 2288–2294.

Cartea, M. E., Francisco, M., Soengas, P., and Velasco, P. (2010). Phenolic compounds in Brassica vegetables. *Molecules*, *16*(1), 251–280.

Chávez-Jáuregui, R. N., Silva, M. E. M. P., and Arĕas, J. A. G. (2000). Extrusion cooking process for amaranth (Amaranthus caudatus L.). *Journal of Food Science*, *65*(6), 1009–1015.

Chen, L., Xin, X., Yuan, Q., Su, D., and Liu, W. (2014). Phytochemical properties and antioxidant capacities of various colored berries. *Journal of the Science of Food and Agriculture*, *94*, 180–188. doi: 10.1002/jsfa.6216

Christaki, E. (2012). Hippophae rhamnoides L. (Sea Buckthorn): a potential source of nutraceuticals. *Food Public Health*, *2*(3), 69–72.

Cláudia, A., Correia-Branco, A., Araújo, J. R., Guimarães, J. T., Elisa, K., and Fátima, M. (2015). The chemopreventive effect of the dietary compound kaempferol on the MCF-7 human breast cancer cell line is dependent on inhibition of glucose cellular uptake. *Nutrition and Cancer*, 1–10. doi: 10.1080/01635581.2015.1002625

Conforti, F., Statti, G., Loizzo, M., Sachetti, G., and Poli, F. (2005). In vitro antioxidant effect and inhibition of a-amylase of two varieties of Amaranthus caudatus seeds. *Biological Pharmaceutical Bulletin*, *28*(6), 1098–1102.

Croft, K. D (1998). The chemistry and biological effects of flavonoids and phenolic acids. *Annals of the New York Academy of Sciences*, *854*, 435–442.

Cuadrado, C., Takacs, K., Szabo, E. E., and Pedrosa, M. (2019). Nonnutritional factors: lectins, phytic acid, proteases inhibitors, allergens. In *Legumes: nutritional quality, processing and potential health benefits*, ed. M. A. Martın-Cabrejas. London: Royal Society of Chemistry. pp. 152–176.

Das, S. (2012). Domestication, phylogeny and taxonomic delimitation in underutilized grain Amaranthus (Amaranthaceae)–a status review. *Feddes Repertorium*, *123*(4), 273–282.

Davis, A. M. (1982). Crude protein, crude fiber, tannin, and oxalate concentrations of 33 Astragalus species. *Rangeland Ecology & Management/Journal of Range Management Archives*, *35*(1), 32–34.

de la Hey, M. and Beinart, W. (2017). Why have South African smallholders largely abandoned arable production in fields? A case study. *Journal of Southern African Studies*, *43*(4), 753–770.

Dlamini, N., Moroka, T., Mlotshwa, L., Reddy, J., and Botha, G. (2010). Indigenous edible plants as sources of nutrients and health benefitting components (nutraceuticals). In *CSIR 3rd Biennial Conference: 2010 Science Real and, Relevant*, CSIR International Convention Center, Pretoria, South Africa, 30 August-1 September 2010. pp. 11.

Dobos, G. (1992). *Koerner amaranth als neve kulturpflanze in Oesterrcich. Introduction and Zuechterische Aspeke*. PhD thesis, University of Natural Resources and Applied Life Sciences.

Dubey, A., Tiwari, D., Singh, Y., Prakash, O., and Singh, P. (2021). Drug repurposing in Oncology: opportunities and challenges. *International Journal of Allied Medical Sciences and Clinical Research*, *9*(1), 68–87.

Ebert, A. W., Wu, T. H., and Wang, S. T. (2011). *Vegetable amaranth (Amaranthus L.)*. AVRDC Publication. No. 11-754. Asian Vegetable Research and Development Center, Shanhua.

Emire, S. A. and Arega, M. (2012). Value added product development and quality characterization of amaranth (Amaranthus caudatus L.) grown in East Africa. *African Journal of Food Science and Technology*, *3*(6), 129–141.

Escudero, N. L., De Arellano, M. L., Luco, J. M., Giménez, M. S., and Mucciarelli, S. I. (2004). Comparison of the chemical composition and nutritional value of Amaranthus cruentus flour and its protein concentrate. *Plant Foods for Human Nutrition*, *59*(1), 15–21.

Fatima, T., Snyder, C. L., Schroeder, W. R., Cram, D., Datla, R., Wishart, D., and Krishna, P. (2012). Fatty acid composition of developing sea buckthorn (Hippophae rhamnoides L.) berry and the transcriptome of the mature seed. *PLoS One*, *7*(4), e34099.

Gandhi, P., Khan, Z., and Niraj, K. (2011). In vitro assay of anti-proliferative potential of Amaranthus cruentus aqueous extract on human peripheral blood lymphocytes. *Current Trends in Biotechnology and Chemical Research*, *1*, 42–8.

Guo, R., Guo, X., Li, T., Fu, X., and Liu, R. (2017). Comparative assessment of phytochemical profiles, antioxidant and antiproliferative activities of Sea buckthorn (Hippophae rhamnoides L.). *Food Chemistry*, *221*, 997–1003. doi: 10.1016/j.foodchem.2016.11.063

Hoidal, N., Jacobsen, S. E., Odone, A., and Alandia, G. (2020). Defoliation timing for optimal leaf nutrition in dual-use amaranth production systems. *Journal of the Science of Food and Agriculture*, *100*(13), 4745–4755.

Holland, T. M., Agarwal, P., Wang, Y., Leurgans, S. E., Bennett, D. A., Booth, S. L., and Morris, M. C. (2020). Dietary flavonols and risk of Alzheimer dementia. *Neurology*, *94*(16), e1749–e1756.

Hu, S. (2005). Hippophae rhamnoides L. – Sea buckthorn, shaji. In *Food plants of China*. The Chinese University of Hong Kong Press, Hong Kong. p. 572.

Jaenicke, H. and Hoeschle-Zeledon, I. eds. (2006). *Strategic framework for underutilized plant species research and development, with special reference to Asia and the Pacific, and to Sub-Saharan Africa*. International Centre for Underutilised Crops, Colombo, Sri Lanka and Global Facilitation Unit for Underutilized Species, Rome, Italy. 33 pp.

Jhade, D., Ahirwar, D., Jain, R., Sharma, N. K., and Gupta, S. (2011). Pharmacognostic standardization, physico-and phytochemical evaluation of Amaranthus spinosus Linn. root. *Journal of Young Pharmacists*, *3*(3), 221–225.

Joshua, L. S., Pal, V. C., Kumar, K. S., Sahu, R. K., and Roy, A. (2010). Antitumor activity of the ethanol extract of Amaranthus spinosus leaves against EAC bearing swiss albino mice. *Der Pharmacia Lettre*, *2*(2), 10–15.

Kalinova, J. and Dadakova, E. (2009). Rutin and Total Quercetin Content in Amaranth (*Amaranthus* spp.). *Plant Foods for Human Nutrition*, *64*, 68–74. doi: 10.1007/s11130-008-0104-x

Karamać, M., Gai, F., Longato, E., Meineri, G., Janiak, M. A., Amarowicz, R., and Peiretti, P. G. (2019). Antioxidant activity and phenolic composition of amaranth (Amaranthus caudatus) during plant growth. *Antioxidants*, *8*(6), 173.

Khan, B. A., Akhtar, N., and Mahmood, T. (2010). A comprehensive review of a magic plant Hippophae rhamnoides. *Pharmacognosy Journal*, *16*, 58–61.

Khanam, U. K. S. and Oba, S. (2013). Bioactive substances in leaves of two amaranth species, Amaranthus tricolor and A. hypochondriacus. *Canadian Journal of Plant Science*, *93*(1), 47–58.

Khushnuma, R., Dakshina, G., Abinav, P. M., and Dubey, A. (2021). Evaluation of hypoglycemic potential of β Escin. *Annals of the Romanian Society for Cell Biology, 25*(6), 13965– 13975. Retrieved from http://annalsofrscb.ro/index.php/journal/article/view/8259

Kim, S., Hwang, E., Yi, S., Song, K., Lee, H., Heo, T., Park, S. K., Jung, Y. J., and Jun, H. S. (2017). Sea buckthorn leaf extracts inhibits glioma cell growth by reducing reactive oxygen species and promoting apoptosis. *Applied Biochemistry and Biotechnology, 182*, 1663–1674. doi: 10.1007/s12010-017-2425-4

Ko, J. H., Sethi, G., Um, J. Y., Shanmugam, M. K., Arfuso, F., and Kumar, A. (2017). The role of resveratrol in cancer therapy. *International Journal of Molecular Sciences, 18*, 2589. doi: 10.3390/ijms18122589

Kolawole, E. L. and Sarah, O. A. (2009). Growth and yield performance of Amaranthus cruentus influenced by planting density and poultry manure application. *Notulae Botanicae Horti Agrobotanici ClujNapoca, 37*(1), 195–199.

Kristo, A., Klimis-Zacas, D., and Sikaliidis, A. (2016). Protective role of dietary berries in cancer. *Antioxidants, 5*, 37. doi: 10.3390/antiox5040037

Kumar, M. Y., Dutta, R., Prasad, D., and Misra, K. (2011). Subcritical water extraction of antioxidant compounds from Seabuckthorn (Hippophae rhamnoides) leaves for the comparative evaluation of antioxidant activity. Food Chemistry, *127*(3), 1309–1316.

Kumar, M. Y., Tirpude, R. J., Maheshwari, D. T., Bansal, A., and Misra, K. (2013). Antioxidant and antimicrobial properties of phenolic rich fraction of Seabuckthorn (Hippophae rhamnoides L.) leaves in vitro. Food Chemistry, *141*(4), 3443–3450.

Kumar, S. and Sagar, A. (2007). Microbial associates of Hippophae rhamnoides (Seabuckthorn). *The Plant Pathology Journal, 6*, 299–305.

Larbie, C., Appiah-Opong, R., Acheampong, F., Tuffour, I., Uto, T., Yeboah, G. A., and Inkabi, S. E. (2015). Anti-proliferative effect of Ficus pumila Linn. on human leukemic cell lines. *International Journal of Basic & Clinical Pharmacology, 4*(2), 330–336.

Leonov, A., Arlia-Ciommo, A., Piano, A., Lutchman, V., Svistkova, V., Medkour, Y., and Titorenko, V. I. (2015). Longevity extension by phytochemicals. *Molecules, 20*, 6544–6572.

Li, H., Deng, Z., Liu, R., Zhu, H., Draves, J., Marcone, M., and Tsao, R. (2015). Characterization of phenolics, betacyanins and antioxidant activities of the seed, leaf, sprout, flower and stalk extracts of three Amaranthus species. *Journal of Food Composition and Analysis, 37*, 75–81.

Li, R., Wang, Q., Zhao, M., Yang, P., Hu, X., and Ouyang, D. (2019). Flavonoid glycosides from seeds of Hippophae rhamnoides subsp. Sinensis with α-glucosidase inhibition activity. *Fitoterapia, 137*, 104248.

Li, T. S. and Beveridge, T. H. (2003). Sea buckthorn (Hippophae rhamnoides L.): production and utilization (No. 45317). Canadian Science Publishing (NRC Research Press), Canada.

Liu, H., Zhang, W., Dong, S., Song, L., Zhao, S., Wu, C., and Wang, Y. (2015). Protective effects of sea buckthorn polysaccharide extracts against LPS/d-GalN-induced acute liver failure in mice via suppressing TLR4-NF-κB signaling. *Journal of Ethnopharmacology, 176*, 69–78.

Maas, J. L., Galletta, G. J., and Stoner, G. D. (1991). Ellagic acid, an anticarcinogen in fruits, especially in strawberries: a review. *Hort Science, 26*(1), 10–14.

Mabhaudhi, T., Chimonyo, V. G. P., Hlahla, S., Massawe, F., Mayes, S., Nhamo, L., and Modi, A. T. (2019). Prospects of orphan crops in climate change. *Planta, 250*(3), 695–708.

Maheshwari, D. T., Yogendra Kumar, M. S., Verma, S. K., Singh, V. K., and Singh, S. N. (2011). Antioxidant and hepatoprotective activities of phenolic rich fraction of sea buckthorn (Hippophae rhamnoides L.) leaves. *Food and Chemical Toxicology, 49*, 2422–2428.

Maldonado-Cervantes, E., Jeong, H. J., León-Galván, F., Barrera-Pacheco, A., De León-Rodríguez, A., De Mejia, E. G., and De La Rosa, A. P. B. (2010). Amaranth lunasin-like peptide internalizes into the cell nucleus and inhibits chemical carcinogen-induced transformation of NIH-3T3 cells. *Peptides*, *31*(9), 1635–1642.

Martínez-Núñez, M., Ruiz-Rivas, M., Vera-Hernández, P. F., Bernal-Muñoz, R., Luna-Suárez, S., and Rosas-Cárdenas, F. F. (2019). The phenological growth stages of different amaranth species grown in restricted spaces based in BBCH code. *South African Journal of Botany*, *124*, 436–443.

Massawe, F., Mayes, S., and Cheng, A. (2016). Crop diversity: an unexploited treasure trove for food security. *Trends in Plant Science*, *21*(5), 365–368.

Meher, C. P., Purohit, D., Kumar, A., Singh, R., and Dubey, A. (2022). An updated review on morpholine derivatives with their pharmacological actions. *International Journal of Health Sciences*, *6*(S3), 2218–2249. doi: 10.53730/ijhs.v6nS3.5983.

Michel, T., Destandau, E., Le Floch, G., Lucchesi, M. E., and Elfakir, C. (2012). Antimicrobial, antioxidant and phytochemical investigations of sea buckthorn (Hippophaë rhamnoides L.) leaf, stem, root and seed. *Food Chemistry*, *131*(3), 754–760.

Miraj, S. (2016) A Medicinal plant with antioxidant activity in Iranian folk medicine: Amaranthus. *Der Pharmacia Lettre*, *8*(14), 27–30.

Murya, N. K. and Pratibha, A. (2018). Amaranthus grain nutritional benefits: a review. *Journal of Pharmacognosy and Phytochemistry*, *7*(2), 2258–2262.

Musa, A., Oladiran, J. A., Ezenwa, M. I., Akanya, M. O., and Ogbadoyi, E. O. (2011). The effects of applied nitrogen fertilizer and leaf positions on levels of micronutrients, anti-nutrients and toxic substances in Amaranthus cruentus. *African Journal of Biotechnology*, *10*(48), 9857–9863.

Nana, F. W., Hilou, A., Millogo, J. F., and Nacoulma, O. G. (2012). Phytochemical Composition, Antioxidant and Xanthine Oxidase Inhibitory Activities of *Amaranthus cruentus* L. and *Amaranthus hybridus* L. Extracts. *Pharmaceuticals*, *5*, 613–628. doi: 10.3390/ph5060613

Neugart, S., Baldermann, S., Ngwene, B., Wesonga, J., and Schreiner, M. (2017). Indigenous leafy vegetables of Eastern Africa—A source of extraordinary secondary plant metabolites. *Food Research International*, *100*, 411–422.

O'Brien, G. K. and Price, M. L. (1983). *Amaranth: grain & vegetable type*. Echo technical note. Durrance Rd, North Ft. Myers, FL.

Olaniyi, J. O. (2007). Evaluation of yield and quality performance of grain amaranth varieties in the southwestern Nigeria. *International Journal of Research in Agronomy*, *1*(2), 42–52.

Olawoye, B. T. and Gbadamosi, S. O. (2017). Effect of different treatments on in vitro protein digestibility, antinutrients, antioxidant properties and mineral composition of Amaranthus viridis seed. *Cogent Food & Agriculture*, *3*(1), 1296402.

Olsson, M., Gustavsson, K., Andersson, S., Nilsson, A., and Duan, R. (2004). Inhibition of cancer cell proliferation in vitro by fruit and berry extracts and correlations with antioxidant levels. *Journal of Agricultural and Food Chemistry*, *52*, 7264–7271. doi: 10.1021/jf030479p

Padulosi, S., Hoeschle-Zeledon, I., and Bordoni, P. (2007). Minor crops and underutilized species: lessons and prospects. In Maxted N, Ford-Lloyd BV, Kell SP, Iriondo JM, Dulloo ME, Turok J (eds) *Crop wild relative conservation and use*. Wallingford, UK: CABI. pp. 605–624.

Patnaik, S., Purohit, D., Biswasroy, P., Diab, W. M., and Dubey, A. (2022). Recent advances for commedonal acne treatment by employing lipid nanocarriers topically. *International Journal of Health Sciences*, *6*(S8), 180–205. doi: 10.53730/ijhs.v6nS8.9671

Popoola, O. O. (2022). Phenolic compounds composition and in vitro antioxidant activity of Nigerian Amaranthus viridis seed as affected by autoclaving and germination. *Measurement: Food*, *6*, 100028.

Qi, F., Sun, J. H., Yan, J. Q., Li, C. M., and Lv, X. C. (2018). Anti-inflammatory effects of isorhamnetin on LPS-stimulated human gingival fibroblasts by activating Nrf2 signaling pathway. *Microbial Pathogenesis*, *120*, 37–41. doi: 10.1016/j.micpath.2018.04.049

Ramesbabu, A. P., Ganapathy, A. P., Jothiramajayam, M., Surbamani, E., and Sundaramoorthy, B. (2011). Review on curative assets of seabuckthorn. *Journal of Research in Pharmacy*, *4*, 164–166.

Rédei, D., Kúsz, N., Jedlinszki, N., Blazsó, G., Zupkó, I., and Hohmann, J. (2018). Bioactivity-guided investigation of the anti-inflammatory activity of Hippophae rhamnoides fruits. *Planta Medica*, *84*(1), 26–33.

Repo-Carrasco-Valencia, R., Hellström, J. K., Pihlava, J. M., and Mattila, P. H. (2010). Flavonoids and other phenolic compounds in Andean indigenous grains: Quinoa (Chenopodium quinoa), kañiwa (Chenopodium pallidicaule) and kiwicha (Amaranthus caudatus). *Food Chemistry*, *120*(1), 128–133.

Ruan, C. J., Rumpunen, K., and Nybom, H. (2013). Advances in improvement of quality and resistance in a multipurpose crop: sea buckthorn. Critical Reviews in Biotechnology, *33*(2), 126–144.

Schmidt, R. R. and Fedtke, C. (1977). Metamitron activity in tolerant and susceptible plants. *Pesticide Science*, *8*(6), 611–617.

Shanmugam, M., Rane, G., Mathi, K. M., Arfuso, F., Chinnathambi, A., and Zayed, M. E. (2015). The multifaceted role of curcumin in cancer prevention and treatment. *Molecules 20*, 2728–2769. doi: 10.3390/molecules20022728

Solcan, C., Gogu, M., Floristean, V., Oprisan, B., and Solcan, G. (2013). The hepatoprotective effect of sea buckthorn (Hippophae rhamnoides) berries on induced aflatoxin B1 poisoning in chickens. *Poultry Science*, *92*, 966–974.

Steffensen, S. K., Pedersen, H. A., Labouriau, R., Mortensen, A. G., Laursen, B., de Troiani, R. M., and Fomsgaard, I. S. (2011). Variation of polyphenols and betaines in aerial parts of young, field-grown Amaranthus genotypes. *Journal of Agricultural and Food Chemistry*, *59*(22), 12073–12082.

Stintzing, F. C., Kammerer, D., Schieber, A., Adama, H., Nacoulma, O. G., and Carle, R. (2004). Betacyanins and phenolic compounds from Amaranthus spinosus L. and Boerhavia erecta L. *Zeitschrift für Naturforschung C*, *59*(1–2), 1–8.

Suryakumar, G. and Gupta, A. (2011). Medicinal and therapeutic potential of Sea buckthorn (Hippophae rhamnoides L.). *Journal of Ethnopharmacology*, *138*, 268–278. doi: 10.1016/j.jep.2011.09.024

Teng, B., Lu, Y., Wang, Z., Tao, X., and Wei, D. (2006). In vitro anti-tumor activity of isorhamnetin isolated from Hippophaer rhamnoides L. against BEL-7402 cells. *Pharmaceutical Research*, *54*, 186–194. doi: 10.1016/j.phrs.2006.04.007

The World Bank Annual Report. (2018). (English). Washington, DC: World Bank Group. https://documents.worldbank.org/curated/en/630671538158537244/The-World-Bank-Annual-Report-2018

Tkacz, K., Wojdylo, A., Turkiewicz, I. P., Ferreres, F., Moreno, D. A., and Nowicka, P. (2020). UPLC-PDA-Q/TOF-MS profiling of phenolic and carotenoid compounds and their influence on anticholinergic potential for AChE and BuChE inhibition and on-line antioxidant activity of selected Hippophae rhamnoides L. cultivars. *Food Chemistry*, *309*, 125766. doi: 10.1016/j.foodchem.2019.125766

Valíček, P. and Havelka, E. V. (2008). Hippophae rhamnoides (in Czech). Start publishing house, Benešov. ISBN 978-80-86231-44-0, 86 p.

Venskutonis, P. R. and Kraujalis, P. (2013). Nutritional components of amaranth seeds and vegetables: a review on composition, properties, and uses. *Comprehensive Reviews in Food Science and Food Safety*, *12*(4), 381–412.

Wang, H., Ge, H., and Zhi, J. (1989). The components of unsaponifiable matters in sea buckthorn fruit and seed oil. In *Proceedings First International Symposium on Sea Buckthorn*, Xi'an. pp. 81–90.

Wang, Y., Nie, F., Ouyang, J., and Wang, X. (2014). Inhibitory effects of sea buckthorn procyanidins on fatty acid synthase and MDA-MB-231 cells. *Tumor Biology*, *35*, 9563–9569. doi: 10.1007/s13277-014-2233-1

Wu, A., Su, Y., Li, J., Liu, Q., Lu, J., Wei, X., Cian, C., Lai, Y., and Wang, G. (1989). The treatment of chronic cervicitis with Hippophae oil and its suppository (129 cases analysis). In *Proceedings first International Symposium on Sea Buckthorn*. pp. 404–406.

Yasukawa, K., Kitanaka, S., Kawata, K., and Goto, K. (2009). Anti-tumor promoters phenolics and triterpenoid from Hippophae rhamnoides. *Fitoterapia*, *80*, 164–167. doi: 10.1016/j.fitote.2009.01.006

Zhamanbaeva, G., Murzakhmetova, M., Tuleukhanov, S., and Danilenko, M. (2014). Antitumor activity of ethanol extract from Hippophae rhamnoides L. leaves towards human acute myeloid leukemia cells in vitro. *Bulletin of Experimental Biology and Medicine*, *158*, 221–224. doi: 10.1007/s10517-014-2734-3

7 Indian Gooseberry
Role in Cancer Prevention and Treatment

Lubna Shafi and Pervaiz Iqbal
B.S. Abdur Rahman Crescent Institute of Science & Technology, Chennai, India

7.1 INTRODUCTION

Indian gooseberry, scientifically known as *Emblica officinalis*, is a fruit berry belonging to the family Euphorbiaceae. Its other vernacular names are amla in Hindi and German, Mirabolano emblico in Italian, Phyllanthe emblica in French, Emblic myrobalan or Malacca tree in English, An Mole in Chinese, Popok Melaka in Malaysian, Skyu-ru-ra in Tibetan, Mirabolano emblico in Portuguese, and Amba in Nepalese. Indian gooseberry is believed to be the first tree created in the universe, as per Indian mythology. It is indigenous to India, although it is also found in the tropical and subtropical regions of Pakistan, China, Uzbekistan, Southeast Asia, Sri Lanka, and Malaysia. India was the world's third-largest producer of amla in 2017–2018, with 11.93 MT productions. Uttar Pradesh, Madhya Pradesh, and Tamil Nadu are the largest amla-cultivating states, accounting for 75.88% of the total amla output in India. The Indian gooseberry is said to have emerged from tropical Southeast Asia. It grows in dry, hot, deciduous, and damp woods (Khan 2009; Singh et al. 2006; NHB 2018).

7.1.1 Plant Anatomy

Indian gooseberry is a small to medium-sized flowering deciduous tree of the same name with a crooked trunk that grows to around 8–18 m in height. It has spreading branches, and the branchlets do not have a smooth or finely pubescent appearance. The flowers are greenish-yellow in color, with grayish-green bark (peels off in flakes) and simple light green leaves similar to pinnate leaves. The little fruit berries are fleshy, spherical with six vertical stripes of furrows, 1–2.5 cm in diameter, deeply ribbed, and have a sour, bitter, and astringent flavor. The fruits are green when unripe and become light yellow or brick red when mature, with six trigonous seeds (acrid and sweet). The mesocarp is the edible component of the fruit, whereas the endocarp is hard and stony and contains the seed (Yahia 2011; Scartezzini 2000).

DOI: 10.1201/9781003477792-7

7.2 NUTRITIONAL VALUE AND CHEMICAL CONSTITUENTS

Indian gooseberry is well known for its nutritional value and contains natural compounds like moisture (81.2%), vitamin C (ascorbic acid), carbohydrates (14.1%), protein (0.5%), fat (0.1%), mineral matter (0.7%), fiber (3.4%), bulk elements, calcium (0.05%), phosphorous (0.02%), iron (1.2 mg/100 g), and nicotinic acid (0.2 mg/100 g). It has more vitamin C than oranges, tangerines, and lemons and is the second most vitamin C-rich fruit after Barbados cherry (*Malpighia glabra* L.), having approximately 600–700 mg per fruit. It is also high in pectin, a kind of dietary fiber that helps to control bowel movements. Its consumption increases the bioavailability of dietary iron from basic foods such as grains and pulses. The protein content in edible Indian gooseberry fruit tissue is three times that of an apple, while the ascorbic acid concentration is 160 times that of an apple. In addition, the fruit has a greater concentration of most minerals and amino acids than apples. Its fruit ash is high in chromium, zinc, and copper and is thought to be an adaptogenic factor that boosts immunity (Srivasuki 2012; Jain & Khurdiya 2004; Scartezzini et al. 2006; Brun & Schumacher 1994).

The chemical constituents present in Indian gooseberry are polyphenols, tannins, and alkaloids. The fruit contains almost 28% of the tannin found in total plants. Tannin is available in two hydrolyzable forms, i.e., emblicanin A and emblicanin B, both of which are antioxidants. It includes a variety of phytochemical components like terpenoids, alkaloids, flavonoids (rutin and quercetin), and tannins, all of which have been found to have biologically beneficial properties and the majority of which are polyphenols, including ellagic acid, gallic acid, chebulagic acid, chebulinic acid, apigenin, corilagin, quercetin, luteolin, etc. It also has a high concentration of bioactive substances such as trigonelline, quercetin, naringin, catechin, and kaempferol. All chemical constituents are described in Table 7.1 (Pareek et al. 2017; Poltanov et al. 2009; Rajak et al. 2004; Sidhu et al. 2011; Arora et al. 2003; Yadav et al. 2017a,b).

TABLE 7.1
Chemical Constituents Present in Indian Gooseberry

S. No.	Bioactive Compounds	Chemical Constituents
1	Phenolic compounds	Gallic acid, ellagic acid, methyl gallate, and trigalloyl glucose
2	Hydrolyzable tannins	Emblicanin A and B, punigluconin, pedunculagin, chebulinic acid, chebulagic acid, corilagin, geraniin, and ellagitannin
3	Flavonoids	Quercetin and kaempferol
4	Alkaloids	Phyllanthin, phyllemblin, and phyllantidine
5	Amino acids	Glutamic acid, proline, aspartic acid, alanine, cystine, and lysine
6	Organic acids	Citric acid
7	Carbohydrates	Pectin
8	Vitamins	Ascorbic acid

TABLE 7.2
Chemical Constituents Present in Different Parts of Indian Gooseberry and its Traditional Uses

Part	Chemical Constituent	Medicinal Uses
Fruit	Protein, carbohydrate, mineral matter, nicotionic acid, fat, fiber, phyllemblin, phyllemblic acid, gallic acid, emblicol, Putranjiva, ellagic acid, pectin, and quercetin	Diarrhea, acute bacillary dysentery, respiratory problems, gout, hair growth and graying prevention, diabetes, constipation, piles, and as a tonic for good health and youthful appearance
Seed	Proteins, fats, mineral matter, fiber, carbohydrate, nicotionic acid (0.2 mg/100 g), phyllemblin, phyllemblic acid, gallic acid, emblico ellagic acid, and pectin It also contains a fixed oil (brownish-yellow in color), phosphatides, and an essential oil	Asthma, biliousness, bronchitis, fever, eye disorder, and nose bleeding
Bark	Tannins, leucodelphinidin	Leucorrhea, gonorrhea, and anti-diarrheic
Leaves	Tannins, gallic acid, ellagic acid, chebulinic acids, alkaloids phyllantidine, and phyllanthin	Dental problems
Root	Ellagic acid and lupeol	Dental problems

All parts of Indian gooseberry, including seeds, fruits, leaves, roots, flowers, and bark, are rich in chemical constituents and have been utilized for medicinal purposes, as described in Table 7.2 (Thakur et al. 1989; Zhang et al. 2004).

7.3 TRADITIONAL USES OF INDIAN GOOSEBERRY

Indian gooseberry is used to treat a range of diseases in Ayurveda, Siddha, and Unani systems of medicine in India, as well as in Tibetan, Sri Lankan, and Chinese systems of medicine. It has a sour, astringent, bitter, sweet, and pungent taste (rasa).

Its properties are as follows:

- Cooling nature (used in treatments for a burning sensation anywhere in the body and for many types of inflammation and fever),
- As a vermifuge,
- Enhances digestion,
- Sharpens the intellect and mental functioning,
- Strengthens the lungs,
- Promotes healthier hair,
- Enhances the glow and luster of the skin,
- For hiccups and for painful respiration,
- For anemia, jaundice, and dyspepsia,
- For hemorrhage, diarrhea, and dysentery.

Indian gooseberry has a long history as a medicinal remedy to cure ulcers and dyspepsia, common cold, and fever, and as a diuretic, refrigerant, laxative, ophthalmic, restorative, liver tonic, stomachic, and anti-pyretic. It is effective in treating diabetes, bronchitis, ophthalmopathy, cough, asthma, cephalalgia, dyspepsia, hyperacidity, colic, flatulence, peptic ulcer, erysipelas, anemia, skin diseases, leprosy, hematogenesis, emaciation, inflammations, jaundice, hepatopathy and diarrhea (Singh et al. 2012; Dasaroju & Gottumukkala 2014; Variya et al. 2016; Diwan et al. 2018; Acharya et al. 2021).

7.3.1 Medicinal Uses of Indian Gooseberry

Indian gooseberry (Figure 7.1) has been used in Ayurveda for decades. According to pharmacological studies, it possesses analgesic, adaptogenic, anti-tussive, gastro, anti-atherogenic, cardiac, neuroprotective, nephrotic, and anti-cancer characteristics. Indian gooseberry is also said to have anti-microbial, hypolipidemic, hypoglycemic, antioxidant, anti-mutagenic, free radical scavenging, chemopreventive, and immune-modulator properties. These properties are useful in the prevention and treatment of degenerative diseases, including cancer, diabetes, liver problems, ulcers, anemia, heart disease, inflammation, asthma, skin problems, neurodegenerative disorders, and obesity (Lim et al. 2016; Nosal'ova et al. 2003; Baliga et al. 2013; Chatterjee et al. 2010; Yokozawa et al. 2007; Madhuri et al. 2011; Jeevangi et al. 2013; Kulkarni & Ghurghure 2018; Vasant & Narasimhacharya 2012; Almatroodi et al. 2020).

The whole plant of Indian gooseberry is useful because of its medicinal and pharmaceutical properties, described in Table 7.2 and Figure 7.2 (Yadav et al. 2017a,b; Jeevangi et al. 2013; Hasan et al. 2016; Baliga & Dsouza 2011; Priya & Islam 2019; Bhide & Nitave 2014; Vasant et al. 2013; Mirunalini & Krishnaveni 2010).

FIGURE 7.1 Indian gooseberry (*Emblica officinalis*).

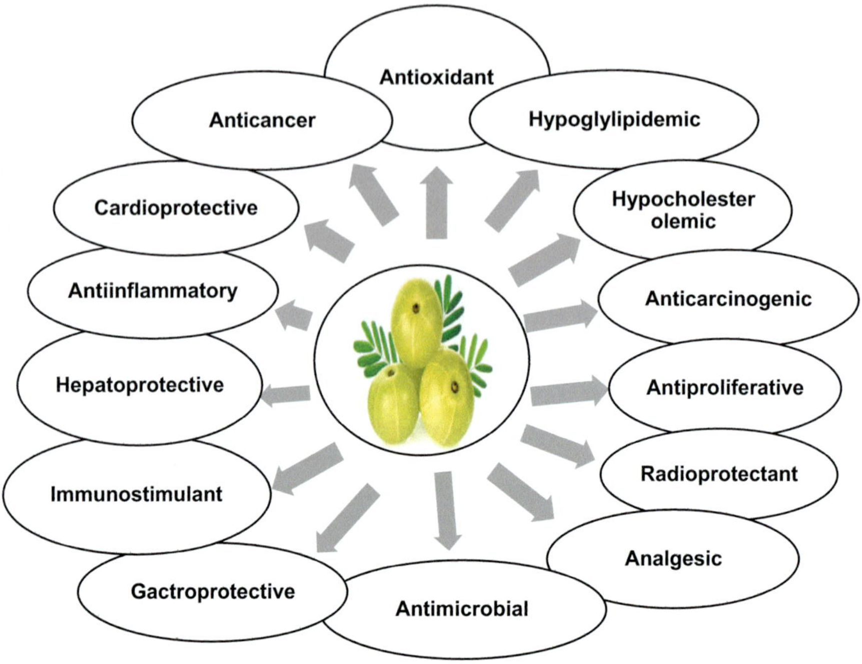

FIGURE 7.2 Various therapeutic applications of Indian gooseberry.

This chapter focuses on Indian gooseberry, known for its nutritional and therapeutic benefits, highlighting its usage in cancer.

7.4 INDIAN GOOSEBERRY IN CANCER

Cancer was first discovered in ancient Egyptian mummies and literature going back to 3000 B.C. The Greek physician Hippocrates (460–370 BC) coined the term "cancer" to characterize carcinoma tumors (Sudhakar 2009). Cancer is a complicated disease characterized by various temporal and spatial alterations in cell physiology that eventually result in malignant tumors. The major cause of morbidity and mortality for most cancer patients is tumor cell invasion of surrounding tissues and distant organs (Seyfried & Shelton 2010). It affects people of all races, gender identities, and socioeconomic backgrounds and is globally the major cause of morbidity and mortality, accounting for 19.3 million new cancer cases and approximately 10 million fatalities (nearly 1 in every 6 deaths) in 2020. The global burden of cancer incidence and mortality is quickly increasing, reflecting both population aging and growth, along with changes in the prevalence and distribution of the key cancer risk factors, many of which are linked to socioeconomic development. An estimated 28.4 million new cancer cases (excluding basal cell carcinoma) are estimated to occur worldwide in 2040, i.e., a 47% rise from 2020 (19.3 million cases). And rises in cancer incidence will definitely be accompanied by a rise in cancer fatality rates. The problem is worse in underdeveloped nations due to a lack of knowledge and insufficient cancer management resources. This disease is a major global health issue, regardless of all advances in medical science (Sung et al. 2021; Siegel et al. 2022; Omran 1971; Balducci & Ershler 2005; López-Gómez et al. 2013).

Cancer is caused by both internal factors (5%–10%) like hereditary mutations, hormones, and immunological disorders, as well as environmental/acquired factors (90%–95%), including tobacco use, unhealthy food, exposure to hazardous radiation, lack of physical exercise, and infectious organisms. The World Health Organization (WHO) 2021 reports revealed that 30%–50% of cancers can be prevented, as certain environmental risk factors, such as eating a healthy diet, being physically active and fit, avoiding cigarettes, and radiation protection, are potentially manageable, implying that cancer can be avoided or has a lower likelihood of developing (Zhang et al. 2020; Wu et al. 2016; Anand et al. 2008; and Colditz et al. 2012). There have also been a number of cancer treatments, such as chemotherapy, radiotherapy, immunotherapy, and surgery, but unfortunately, there are associated complications like myelotoxicity, cardiotoxicity, nephrotoxicity, neurotoxicity, and immunosuppression, as well as financial stress. Treatments such as chemotherapy and ionizing radiation are linked to negative side effects because their cytotoxic effects are unbiased, and when combined with neoplastic cells, they can also harm normal tissues. In light of these findings, researchers concentrated their efforts on the hunt for anti-cancer compounds/products in natural products derived from medicinal plants/herbal medicine (Cuzick 2017; Smith et al. 2019; Cragg et al. 2009).

Medicinal plants have been used in traditional medicine across the world to cure a range of diseases for ages. Currently, 25%–28% of the medications used in modern medicine are derived directly or indirectly from plant metabolites. There has also been progress in the development of plant-based medications for the treatment and prevention of cancer. Medicinal plants are significant for cancer prevention and treatment because they include a variety of chemical constituents that may be used to identify novel cancer-fighting compounds. There are a variety of medicinal plants all over the world, a few of which are *Curcuma longa* (turmeric), *Zingiber officinale* (ginger), *Moringa oleifera*, *Azadirachta indica* (Neem), *E. officinalis* (Indian gooseberry), *Trigonella foenum-graecum* (fenugreek), *Papaver somniferum* (opium), and so on. In this chapter, the focus will be on the anti-cancer properties of Indian gooseberry and its role in cancer prevention and treatment (Fridlender et al. 2015; Ortega & Campos 2019; Mir et al. 2019).

7.5 MECHANISM OF ANTI-CANCER ACTIVITY OF INDIAN GOOSEBERRY

Indian gooseberries are a good source of vitamin C and low-molecular-weight hydrolyzable tannins, making them an excellent antioxidant. They are high in bioactive substances such as phenolics, flavonoids, and tannins, and these phytochemicals have been shown to have anti-cancer and antioxidant properties. Anti-cancer effects can be partly ascribed to the antioxidant action mediated by the fruit's various polyphenolic compounds.

Indian gooseberry contains phenolic compounds and tannins, which can efficiently scavenge free radicals in a concentration-dependent manner, protecting the body from oxidative stress. The antioxidant process includes the scavenging of reactive oxygen species that can induce oxidative damage to cellular macromolecules such as DNA, RNA, proteins, and lipids. Thus, the buildup of oxidative DNA damage promotes tumor growth, and an increasing amount of information supports the fact that oxidative stress is one of the key reasons for promoting carcinogenesis. The antioxidant activity of the phenolic compounds of Indian gooseberry was attributable mostly to the phenolic content, with gallic acid and tannic acid found as the principal antioxidant components in the phenolic fractions of this fruit. It has been reported that the antioxidant properties of Indian gooseberries are directly related to their anti-cancer potential. The mechanism of the Indian gooseberry extract's anti-cancer impact on cervical cancer cells has been thoroughly studied. It induces total and partial abrogation of constituent components of the DNA-binding proteins, like c-Jun, JunB, JunD, and c-Fos, as well as down-regulates the expression of the activator protein-1 (AP-1) component and other proteins regulated by AP-1 in the redox transcriptional apparatus. All of these impacts on the AP-1 signaling pathway resulted in a dose-dependent reduction of human papillomavirus-16 (HPV16) and HPV18 transcription, which inhibited viral oncogene production. It has been revealed that ascorbic acid, a primary component of Indian gooseberry, also prevents cancer (Sosa et al. 2013; Mahata et al. 2013; Juranić & Žižak 2005; Frei 1994; Baby et al; 2018).

7.5.1 Prevention of Cancer by Phytochemicals of Indian Gooseberry

Phytochemicals, also known as secondary metabolites, are a class of substances that plants produce in response to environmental signals, pathogen infections, or nutrient deficiency. Indian gooseberry phytoconstituents work together to make it a powerful free radical scavenger, protecting DNA from reactive oxygen species damage. It is high in polyphenols and hydrolyzable tannin-derived substances, which have been shown to inhibit carcinogen-induced mutagenesis and lipid peroxidation. The majority of the active ingredients in Indian gooseberry such as tannins, chebulagic acid, ellagic acid, and others exhibit anti-proliferative and pro-apoptotic properties in cancer cells. The potential of polyphenols to behave as antioxidants or scavengers of reactive oxygen species has been directly linked to their favorable effect on cancer prevention and treatment. They possess anti-cancer effects by targeting several cellular activities and numerous signaling pathways, and they have been shown to stimulate the expression of genes encoding antioxidant and detoxifying enzymes. The majority of them have also exhibited anti-proliferative effects on tumor cells as well as improved cancer cell sensitivity to conventional chemotherapeutic drugs and radiation (Makena & Chung 2007; Ismail et al. 2016; Afrin et al. 2016). The Indian gooseberry fruit extract has the ability to prevent dreadful diseases by assisting the body in combating the negative impacts of carcinogenic substances and inhibiting the further proliferation of affected cells (Zhang et al. 2004; Khan et al. 2019). It was also revealed that it had the greatest effect on the stomach, skin, and uterus.

Indian gooseberry and its phytochemicals (gallic acid, ellagic acid, pentagalloylglucose, quercetin, pyrogallol, and kaempferol) have been shown to be cytotoxic to cancerous cells while normal cells remain unaffected. It's likely that these substances have an effect on neoplastic cells that go through an abnormal cell cycle. These chemicals appear to cause apoptosis and cytotoxicity by altering the proteins involved in cell development. Nor-sesquiterpenoid glycosides derived from the roots have anti-proliferative properties. Quercetin has been shown to promote apoptosis by blocking growth factor signaling pathways like Estimated Glomerular Filtration Rate (EGFR) and thereby modifying the relay mechanism of oncogenic growth signals (Zhao et al. 2015).

7.5.1.1 Tannins

Tannins are naturally occurring polyphenolic compounds with a relatively high molecular weight. It has anti-carcinogenic and anti-mutagenic properties that may be connected to its anti-oxidative activity, which is vital in protecting cells from oxidative damage, particularly lipid peroxidation. Tannins, including chebulagic acid, ellagic acid, and corilagin, inhibit Nuclear Factor kappa B (NF-kappaB) and have all been found to have anti-proliferative and pro-apoptotic actions against cancer cells. The hydrolyzable tannins in Indian gooseberry have also been shown to have specific cytotoxicity to human oral squamous cell carcinoma and salivary gland tumor cell lines, while human gingival fibroblasts remain unaffected. Ellagic acid has been shown to have an anti-carcinogenic impact against breast cancer and prostate adenocarcinoma in animal models. Gallic acid protects against osteosarcoma by triggering apoptosis in tumor cells. Chebulagic acid, ellagic acid, corilagin, and gallic acid are the main mediators of NF-K inhibition and eventually cause cancer cells to undergo apoptosis. Corilagin's cytotoxic effects on ovarian cancer cell lines SKOv3ip, Hey, and HO-8910PM indicated induction of G2/M cell cycle arrest and hence programmed cell death by targeting the Transforming Growth Factor, Protein Kinase B, Extracellular Signal-Regulated Kinase and Suppressor of Mothers against Decapentaplegic (TGF-/AKT/ERK/Smad) signaling pathways. Chebulagic acid, the primary component of Indian gooseberry, is a benzopyran tannin compound that promotes autophagy (Chung et al. 1998; Vadhanam et al. 2011; Edderkaoui et al. 2008; Anitha et al. 2013; Naiki-Ito et al. 2015; Liang et al. 2012; Yıldırım & Kutlu T 2015; Kim et al. 2014).

7.5.1.2 Flavonoids

Flavonoids are polyphenolic compounds with anti-cancer properties that include modulating reactive oxygen species-scavenging enzyme activities, participating in cell cycle arrest, inducing apoptosis and autophagy, and suppressing cancer cell proliferation and invasiveness. Indian gooseberry contains a high concentration of flavonoids such as catechin, quercetin, naringin, trigonelline, and kaempferol, which promote anti-proliferative activities by influencing the Wnt/β-catenin signaling pathway at multiple levels, ranging from ligand-receptor interaction to methylation of genes expressing pathway components. Vadde et al. (2016) explored the bioactive profile of Indian gooseberry extracts as well as their anti-proliferative and pro-apoptotic effects on human colon cancer stem cells. The findings suggest that

Indian gooseberry extracts (seed and pulp) inhibit human colon cancer stem cell proliferation and promote apoptosis regardless of p53 status by likely targeting the Wnt/β-catenin signaling pathway (Tarapore et al. 2012; Amado et al. 2014; Vadde et al. 2016).

Indian gooseberry has been demonstrated to have anti-cancer properties in preclinical research using both in vitro and in vivo systems. Sharma et al. (2000a, 2000b) used the Ames test (a standard measure of carcinogenicity) to evaluate the effectiveness of aqueous extracts (containing all ingredients) of Indian gooseberry containing vitamin C in reducing the mutagenicity of aflatoxin B_1 (a potent dietary carcinogen) and benzo(a)pyrene (an environmental carcinogen). It was confirmed that Indian gooseberry inhibits mutagenesis in vitro. Jose et al. (2001) found that Indian gooseberry extract effectively inhibited the development of L929 cells in culture and considerably decreased solid tumors caused by Dalton's lymphoma ascites (DLA) cells while having only a minor effect on ascites tumors.

Mahata et al. (2013) studied the mechanism of Indian gooseberry fruit extract by examining its influence on AP-1 activity and HPV transcription, both of which are required for cervical cancer cells to be tumorigenic. Indian gooseberry inhibited the DNA-binding activity of constitutively active AP-1 in both HPV16-positive and HPV18-positive cervical cancer cells in a dose- and time-dependent manner. It shows anti-cancer properties by inhibiting AP-1 and suppressing viral transcription, resulting in growth inhibition in cervical cancer cells. Its growth inhibitory action was predominantly expressed by the induction of cell apoptosis. These findings demonstrate that Indian gooseberry inhibits AP-1 and targets transcription of viral oncogenes involved in cervical cancer formation and progression, implying that it might be used to treat HPV-related cervical malignancies.

The majority of in vitro studies showed that the Indian gooseberry extracts have an anti-proliferative impact that is mediated by the activation of an apoptotic cascade. It's noteworthy that either purified or crude Indian gooseberry extracts can inhibit NF-kappaB. In vitro, the anti-proliferative behavior of medicinal plant extracts against human tumor cell lines, including human erythromyeloid K562, T-lymphoid Jurkat, B-lymphoid Raji, and erythroleukemia HEL cell lines, was compared. Pyrogallol, an active compound present in Indian gooseberry, known to cause activation of oxygen species and NF-kappaB binding, was found to be the most effective in inhibiting in vitro cell proliferation. Lung cancer xenograft growth is inhibited by pyrogallol at just 75 μ/kg. Pyrogallol has been shown to suppress the growth of As4.1 JG cells and HeLa cells by cell cycle arrest and apoptosis, suggesting that the substance has anti-proliferative properties in these cells. In response to pyrogallol exposure for 4 hours, lung cancer cell development is halted in the G2/M phase, and the proteins' expressions of cyclin B1, cdc2, and cdc25c are reduced while inactive p-cdc 25c is raised (Ho et al. 2013; Penolazzi et al. 2008; Khan et al. 2002; Yang et al. 2009; Kim et al. 2014). The mechanism of action of Indian gooseberry is shown in Table 7.3.

The discussion presented above clearly indicates that the Indian gooseberry has the ability to inhibit the occurrence of many cancers and is a non-toxic anti-cancer agent.

TABLE 7.3
Mechanism of Action of Indian Gooseberry

Chemical Class	Mechanism
Polyphenols	Apoptosis, anti-proliferative, inhibit mutagenesis
Ellagic acid	Anti-carcinogenic, apoptosis, antioxidant, anti-proliferative, radio-protective, and chemopreventive
Gallic acid	Controls metastasis, radio-protective effect, chemopreventive effect, anti-oxidative, anti-mutagenic, and anti-carcinogenic
Tannins	Anti-proliferative, anti-tumorigenic, and activates autophagy and apoptosis
Quercetin	Promotes apoptosis and potential chemopreventive activity
Corilagin	Anti-proliferative and pro-apoptotic
Pyrogallol	Inhibits cell proliferation
Chebulagic acid	Autophagy, anti-proliferative, and pro-apoptotic
Flavonoids	Apoptosis and autophagy, and suppresses cancer cell proliferation and invasiveness

Therefore, the anti-cancer properties of Indian gooseberry are thought to be due to a number of mechanisms, which are stated below:

- Antioxidant, anti-inflammatory, and anti-mutagenic properties,
- Excessive Phase 1 liver enzymes are reduced (which can convert chemicals into carcinogens),
- Reduced lipid peroxidation,
- Apoptosis induction in neoplastic and preneoplastic cells,
- Rise in glutathione S-transferase (GST) levels, an antioxidant enzyme,
- Defends against DNA damage caused by a variety of carcinogens,
- Prevents metastasis,
- Modulates the levels of proteins important in cell cycle progression.

7.6 INDIAN GOOSEBERRY IN CANCER TREATMENT

Indian gooseberry has been found to have radiomodulatory, chemopreventive, chemomodulatory, immunomodulatory, and anti-mutagenic properties, all of which are effective for the treatment of cancer. In vitro and in vivo studies have shown that Indian gooseberry extracts have significant tumor-suppressive activities against a range of cancer types.

Indian gooseberry extract enhanced with polyphenols or simple aqueous extracts has also revealed cytotoxic activity against cervical and ovarian cancer cells. Its polyphenol extract (150 mg/mL) inhibits HeLa cell growth by causing cell cycle arrest at the G2/M phase and triggering apoptosis (Zhu et al. 2013). Indian gooseberry is a well-known hepatoprotectant. Shivananjappa and Joshi (2011) studied the effects of *E. officinalis* extracts (from dried fruits) on cellular oxidative state using a hepatocyte cell line (HepG2). It was discovered that aqueous extracts of Indian gooseberry boosted endogenous antioxidant defenses in HepG2 cells. Zhang et al. (2004) found that the phenolic compounds and key components from the fruit juice of *E. officinalis*, as well as the branches, leaves, and roots, inhibited B16F10 cell development more effectively than HeLa and MK-1 cell development.

Guo et al. (2013) investigated the effects of Indian gooseberry fruit aqueous extract on cell death and genomic damage in human colon adenocarcinoma cell lines. The results revealed a substantial reduction in the nuclear division index and necrosis in a time- and dose-dependent way. Furthermore, it dramatically elevated apoptosis and demonstrated a strong link between apoptosis and genomic instability (GIN). Guo and Wang's (2016) study looked at the effects of Indian gooseberry on biomarkers related to the spindle assembly checkpoint, mitotic aberrations, and GIN in human NCM460 normal colon epithelial cells. The findings demonstrated that Indian gooseberry has the ability to protect human normal colon epithelial cells from mitotic and genomic damage, partially by improving the function of the spindle assembly checkpoint.

Aqueous extracts of the Indian gooseberry were given orally against an ovarian cancer xenograft model, and it was observed that the extracts were able to totally ablate the development of ovarian cancer xenografts, with little or no residual tumor detected after treatment (De et al. 2013). To investigate the function of anti-metastatic, anti-proliferative, and anti-angiogenic activities of the extract, De et al. (2020) conducted a pioneer study on p53 and BRCA1/2 mutation-bearing, taxol-resistant, and carboplatinum-resistant high-grade serous epithelial ovarian cancer cells. It has been seen that Indian gooseberry extract significantly reduced the migratory and invasiveness features of all investigated high-grade serous epithelial ovarian cancer cell phenotypes; significantly enhanced the expression of E-cadherin; and decreased the expression of IGF1R, SNAIL1, and HIF-1α. Furthermore, mouse xenograft tumor regression obtained from OV4855 cells was seen following oral administration of amla extract. The Indian gooseberry extract targeted SNAIL1 and miR-375, repressed the production of E-cadherin, and reduced the IGF1R gene and protein expression in ovarian cancer cells. Additionally, the extract increased miR-375 levels in exosomes in the medium. Finally, the IGF1R and SNAIL1 proteins were significantly reduced in SKOV3-derived xenograft tumor attenuation extract (De et al. 2016). Yahayo et al. (2013) examined the anti-metastatic potential of an Indian gooseberry aqueous extract in their study. The findings demonstrated that the anti-metastatic activity of *E. officinalis* extract reduces cancer cell invasion, adhesion, migration, and proliferation in a time- and dose-dependent way. MMP2 and MMP9 both showed a reduction in expression, indicating a biological mechanism for anti-metastasis.

Triphala has been shown to have chemopreventive properties, and its inclusion in the diet reduced benzo(a)pyrene-induced forestomach papilloma genesis in mice. When compared to its separate parts, it was more effective in lowering tumor incidence. Triphala is an herbal treatment composed of three dried fruits: *E. officinalis*, *Terminalia bellirica*, and *Terminalia chebula*. Triphala also enhanced the antioxidant state of the mice, possibly contributing to chemoprevention. The cytotoxic effects of Triphala aqueous extract on a transplantable mouse thymic lymphoma and a human breast cancer cell line were examined. The distinct responses of normal and tumor cells to Triphala in vitro, as well as the significant regression of transplanted tumors in mice given Triphala, point to its potential utility as a therapeutic anti-cancer treatment.

It has been shown that gallic acid, a key polyphenol found in "Triphala," inhibits the proliferation of cancer cells. Indian gooseberry boosts the antioxidant and anti-cancer properties of *Semecarpus anacardium* nut milk extract. Indian gooseberry fruit's aqueous extract effectively reduced the growth of six human cancer cell lines, viz., lung, liver, cervical, breast, ovarian, and colorectal. Its cytotoxic action on cancer cells included the production of apoptosis via the extrinsic route, in which caspase-8 and caspase-3/7 were significantly activated and Fas expression increased. The extract has anti-tumor-boosting action, as it reduces tumor incidence in mice. Furthermore, Indian gooseberry extract exhibited potential anti-invasive action against human breast cancer cells (Deep et al. 2005; Sandhya et al. 2006; Kaur et al. 2005; Arulkumaran et al. 2006; Ngamkitidechakul et al. 2010). Generally, Indian gooseberry extract appears to have substantial cytotoxic effects on most cell lines.

7.6.1 Usage of Indian Gooseberry against Chemical Carcinogenesis

Indian gooseberry has been shown to reduce cyclophosphamide-induced DNA damage in mice bone marrow cells, as evaluated by a decrease in both micronuclei and chromosomal aberration. The treatment of 250 and 500 mg/kg of Indian gooseberry extract effectively decreased the genotoxicity of benzo(a)pyrene and cyclophosphamide (both powerful carcinogens) in both test methods. Indian gooseberry lowered cytochrome P450 levels and raised antioxidant glutathione levels, antioxidant enzymes [glutathione peroxidase (GPx), glutathione reductase], and detoxification enzyme GST levels. Its fruit extract drastically reduced the cytotoxic effects of a known carcinogen, 3,4-benzo(a)pyrene, as well as the clastogenicity induced by lead and aluminum in mice. Chemoprevention using dietary phytochemicals is now regarded as one of the most significant cancer-control techniques. Indian gooseberry is appreciated for its specific tannins and flavonoids, which have extremely potent antioxidant qualities. The fruit extract of this plant was tested for tumor suppression in a two-stage process of tumorigenesis in Swiss albino mice. The continuous administration of Indian gooseberry extract at 100 mg/kg to mice decreased tumor incidence by ~60% in a skin carcinogenesis model. Indian gooseberry plant-derived phenolic compounds have a variety of beneficial properties and have the ability to suppress numerous stages of carcinogenesis. Thus, Indian gooseberry or its pharmaceutical formulations may be effective as part of a combination treatment for cancer patients (Sharma et al. 2000a, b; Dhir et al. 1990; Sancheti et al. 2005).

Jeena et al. (1999) in their study found that the hepatocarcinogenesis caused by N-nitrosodiethylamine is effectively prevented by the simultaneous administration of the extracts of Indian gooseberry. It has long been known as a liver tonic and has been widely utilized in traditional medicine. Furthermore, Indian gooseberry polyphenol or aqueous fractions given at 60–250 mg/kg inhibited *N*-nitrosodiethylamine-induced hepatocellular cancer by 80%–100%. The effects of the extract on liver tumor incidence, levels of carcinogen-metabolizing enzymes, levels of cancer indicators, and damage markers were used to assess the anti-carcinogenic activity of the extract. The liver's morphology and marker enzyme levels suggested that the Indian gooseberry extract provided protection against chemical carcinogenesis. Rajeshkumar et al. (2003) examined the effectiveness of the Indian gooseberry polyphenol fraction in inducing apoptosis in mouse and human cancer cell lines, as well as its modulatory influence on N-nitrosodiethylamine-induced liver tumors in rats. In DLA and CeHa cell lines, Indian gooseberry polyphenol therapy induced apoptosis.

Indian gooseberry polyphenol also reduced DNA topoisomerase in *Saccharomyces cerevisiae*, mutant cell cultures, and cdc25 tyrosine phosphatase activity. The outcomes also revealed that Indian gooseberry polyphenol fraction therapy could reduce the development of liver tumors caused by N-nitrosodiethylamine. The ethanolic extract of Indian gooseberry was tested for its ability to protect against dimethylbenz[a]anthracene (DMBA)-induced genotoxicity. The frequency of bone marrow micronuclei was reduced in Swiss albino mice given Indian gooseberry fruit orally at varied dosages (100, 250, and 500 mg/kg b.wt.) for 7 days prior to a single intraperitoneal injection of DMBA. The antioxidant potential of Indian gooseberry, as well as its modulatory influence on hepatic activation and detoxifying enzymes, may contribute to its protection. DMBA is an immunosuppressant and a strong organ-specific laboratory carcinogen, present in tobacco smoke, that is commonly utilized in cancer research labs. Cytochrome, e.g., P450 2E1, is crucial in alcohol and toxin metabolism; it catalyzes the conversion of substrates into more polar metabolites and generates reactive oxygen species. Oxidative hepatocyte stress caused by reactive oxygen species increases damage and death, which leads to inflammation, hepatic stellate cell activation, and liver fibrosis.

Indian gooseberry extract contains properties that allow it to lower the levels of cytochrome enzymes in liver cells. It reduced N-nitrosodiethylamine-induced liver apoptosis and autophagy by modulating the Bax/Bcl-2 ratio and Beclin-1 expression. N-nitrosodimethylamine is a yellow, unscented liquid compound that is a human carcinogen. Hence, Indian gooseberry intake protects the liver against N-nitrosodiethylamine-induced damage through antioxidant, anti-inflammation, anti-apoptosis, and anti-autophagy activities. Sultana et al. (2008) studied the chemopreventive effects of Indian gooseberry on rat hepatocarcinogenesis caused by diethylnitrosamine and 2-acetylaminofluorene. It improved the carcinogenic response by reversing histological alterations and decreasing the number of gamma-GT-positive foci caused by the Solt-Farber regimen in the liver of Wistar rats. It was found that Indian gooseberry has the ability to reduce carcinogen-induced responses in rat liver (Banu et al. 2004; Chen et al. 2011).

7.6.2 Usage of Indian Gooseberry against Radiotherapy

Ionizing radiation has long been used in cancer treatment, and its benefit is enormous when surgery is dangerous or may affect the patient. However, the therapeutic benefit of radiation therapy is countered by radiation damage to neighboring healthy tissues. Therefore, utilizing radio-protective substances that can particularly protect normal tissues from radiation damage is regarded as very beneficial in these situations. The radio-protective properties of Indian gooseberry have been studied in animal models (Kudva et al. 2022). Singh et al. (2005) revealed that giving mice 50, 100, 200, 400, and 800 mg/kg b.wt./day Indian gooseberry once daily for 7 consecutive days before exposing them to a sublethal dosage of gamma radiation prevented them from radiation-induced disease and death. The best response was reported at 100 mg/kg b.wt., which delayed radiation-induced lethality and produced a survival rate of 87.5% when compared to placebo-treated irradiated cohorts where no survivors were recorded. Hari Kumar et al. (2004) found that treating mice with 2.5 g/kg b.wt. of Indian gooseberry for 10 consecutive days before exposing them to a single dose of 7Gy radiation boosted total leukocyte count, bone marrow vitality, and hemoglobin levels. However, therapy with Indian gooseberry after irradiation (continuously for an additional 15 days) was not as successful as administration before radiation, implying that it should be used only when radiation exposure is scheduled.

Indian gooseberry increased the activity of numerous antioxidant enzymes like catalase, superoxide dismutase, and GPx; the phase II detoxifying enzyme, GST; and the antioxidant thiol, glutathione, in the blood, with a concurrent drop in lipid peroxide levels. Jindal et al. (2009) found similar results in the intestines of mice, confirming that Indian gooseberry considerably inhibits the deleterious effects of radiation, at least in part through its antioxidant and lipid peroxidation inhibitory capabilities. Hegde et al. (2022) investigated the effectiveness of Indian gooseberry in reducing radiation-induced mucositis, weight loss, and tumor control when taken with povidone-iodine. During curative radiotherapy (>60 Gy), patients with head and neck cancer were given amla in conjunction with iodine or just iodine. The findings demonstrate that the group that gargled with iodine and Indian gooseberry was very effective in delaying mucositis, reducing the incidence of unbearable mucositis, weight loss quantitative grade, and weight loss severity without compromising tumor response when compared to iodine alone. Triphala has also been shown to be an effective radio-protective agent in mice exposed to radiation. Treating mice with varying doses of Triphala for 5 days before irradiation delayed the onset of fatality and reduced the symptoms of radiation illness when compared to non-drug-treated irradiated controls (Jagetia et al. 2002).

7.6.3 Usage of Indian Gooseberry against Chemotherapy Drugs

Pinmai examined the growth inhibitory effects of Indian gooseberry extracts on human hepatocellular carcinoma and lung carcinoma cells, as well as their synergistic impact with doxorubicin or cisplatin. Doxorubicin and cisplatin are two widely used anti-cancer medicines, which are very harmful to normal tissues during cancer treatment. The results revealed that Indian gooseberry efficiently inhibited the growth of human hepatocellular carcinoma and lung carcinoma cells and integrated the cytotoxic impacts of doxorubicin and cisplatin. The ethanolic extract of Indian gooseberry also safeguarded cardiac myoblast cells from doxorubicin-induced toxicity. These findings point to the possibility that Indian gooseberry protects normal cardiac myoblasts from doxorubicin-induced cardiotoxicity while also sensitizing cancer cells to anti-neoplastic effects. The aqueous extract of Indian gooseberry has been shown to be efficient in lowering cyclophosphamide-induced reduction of humoral immunity and in restoring glutathione and antioxidant enzyme levels in mouse kidneys and liver. Cyclophosphamide is among the most popular alkylating anti-cancer medicines, despite its hazardous side effects, including immunotoxicity, hematotoxicity, mutagenicity, and so on. Immu-21 (a poly-herbal formulation comprising extracts of *Ocimum sanctum*, *Withania somnifera*, *P. emblica*, and *Tinospora cordifolia*) was reported to protect mice against cyclophosphamide-induced genotoxicity (Haque et al. 2001; Pinmai et al. 2008; Wattanapitayakul et al. 2005; Jena et al. 2003).

Rosarin et al. (2013) produced silver Indian gooseberry nano-particles and tested their anti-tumor efficacy in laryngeal carcinoma cells. It was found that the IC50 of Indian gooseberry alone and Indian gooseberry-silver nano-particles were 30 and 20 mg/mL, respectively, and that these nano-particles demonstrated potential anti-proliferative effects through oxidative stress, mitochondrial depolarization, DNA damage, and apoptotic function in laryngeal carcinoma cells, with results comparable to cells treated with 5-FU.

Thus, the significance of Indian gooseberry is that it can be utilized as a natural dietary supplement to combat the cytotoxic effects of mutagens and carcinogens induced through medications, diet, or environment.

7.7 CONCLUSION

Cancer is a serious public health issue and the world's second biggest cause of death, after cardiovascular disease. Although there are different effective cancer treatments available, they all have the potential to cause toxicity and harmful impacts that can lead to major consequences as well as financial stress. As a result, there is a demand for choosing different therapies based on naturally derived anti-cancer drugs, with plants as the preferred source. Natural-derived substances are thought to have fewer or no hazardous side effects than conventional therapies such as chemotherapy and radiotherapy. Medicinal plants are an important part of traditional indigenous healthcare systems since they help to overcome all of these problems. Researchers all around the world have started to realize the significance of traditionally used non-toxic plants for the discovery and production of new medications. Indian gooseberry is the most popular medicinal plant used in Ayurvedic medical products. It is one of the most abundant natural sources of vitamin C and is regarded as a safe herbal treatment with no side effects. For decades, both traditional practitioners and modern medical doctors have utilized it in different crude and refined formulation forms to prevent and treat cancer.

Indian gooseberry, a super-food source of vitamin C (anti-toxin), is useful in the battle against cancer. It can efficiently neutralize or reduce the harmful effects of the majority of chemical carcinogens in food and the environment, making it extremely valuable in cancer prevention programs as well as cancer therapy. Indian gooseberry and some of its phytochemicals, including gallic acid, pyrogallol, ellagic acid, some nor-sesquiterpenoids, corilagin, elaeocarpusin, geraniin, and prodelphinidins B1 and B2, have anti-neoplastic properties. Indian gooseberry also possesses chemomodulatory, radiomodulatory, chemopreventive, antioxidant, free radical scavenging, anti-mutagenic, anti-inflammatory, and immunomodulatory characteristics, all of which are effective in the treatment and prevention of cancer. However, more studies on the efficiency of these phytochemicals/extracts against many types of malignancies, both in vitro and in vivo, are required to develop a comprehensive knowledge of their chemotherapeutic usefulness in cancer prevention. Indian gooseberry remains a hot topic with several possibilities for further research because of its availability, low price, safety in intake, and multiple medical activities.

REFERENCES

Acharya, C. K., Khan, N. S., & Madhu, N. R. (2021). Medicinal uses of Amla, Phyllanthus emblica L.(Gaertn.): a prospective review. *Mukt Shabd Journal*, *10*, 311–315.

Afrin, S., Giampieri, F., Gasparrini, M., Forbes-Hernandez, T. Y., Varela-López, A., Quiles, J. L., & Battino, M. (2016). Chemopreventive and therapeutic effects of edible berries: a focus on colon cancer prevention and treatment. *Molecules*, *21*(2), 169.

Almatroodi, S. A., Alsahli, M. A., Almatroudi, A., Dev, K., Rafat, S., Verma, A. K., & Rahmani, A. H. (2020). Amla (Emblica officinalis): role in health management via controlling various biological activities. *Gene Reports*, *21*, 100820.

Amado, N. G., Predes, D., Moreno, M. M., Carvalho, I. O., Mendes, F. A., & Abreu, J. G. (2014). Flavonoids and Wnt/β-catenin signaling: potential role in colorectal cancer therapies. *International Journal of Molecular Sciences*, *15*(7), 12094–12106.

Anand, P., Kunnumakara, A. B., Sundaram, C., Harikumar, K. B., Tharakan, S. T., Lai, O. S.,... & Aggarwal, B. B. (2008). Cancer is a preventable disease that requires major lifestyle changes. *Pharmaceutical Research*, *25*(9), 2097–2116.

Anitha, P., Priyadarsini, R. V., Kavitha, K., Thiyagarajan, P., & Nagini, S. (2013). Ellagic acid coordinately attenuates Wnt/β-catenin and NF-κB signaling pathways to induce intrinsic apoptosis in an animal model of oral oncogenesis. *European Journal of Nutrition*, *52*(1), 75–84.

Arora, S., Kaur, K., & Kaur, S. (2003). Indian medicinal plants as a reservoir of protective phytochemicals. *Teratogenesis, Carcinogenesis, and Mutagenesis*, *23*(S1), 295–300.

Arulkumaran, S., Ramprasath, V. R., Shanthi, P., & Sachdanandam, P. (2006). Restorative effect of Kalpaamruthaa, an indigenous preparation, on oxidative damage in mammary gland mitochondrial fraction in experimental mammary carcinoma. *Molecular and Cellular Biochemistry*, *291*(1), 77–82.

Baby, B., Antony, P., & Vijayan, R. (2018). Antioxidant and anticancer properties of berries. *Critical Reviews in Food Science and Nutrition*, *58*(15), 2491–2507.

Balducci , L., & Ershler, W. B. (2005). Cancer and ageing: a nexus at several levels. *Nature Reviews Cancer*, *5*(8), 655–662.

Baliga, M. S., & Dsouza, J. J. (2011). Amla (Emblica officinalis Gaertn), a wonder berry in the treatment and prevention of cancer. *European Journal of Cancer Prevention*, *20*(3), 225–239.

Baliga, M. S., Prabhu, A. N., Prabhu, D. A., Shivashankara, A. R., Abraham, A., & Palatty, P. L. (2013). Antidiabetic and cardioprotective effects of Amla (Emblica officinalis Gaertn) and its phytochemicals: preclinical observations. In *Bioactive food as dietary interventions for diabetes* (pp. 583–600). Academic Press.

Banu, S. M., Selvendiran, K., Singh, J. P. V., & Sakthisekaran, D. (2004). Protective effect of Emblica officinalis ethanolic extract against 7, 12-dimethylbenz (a) anthracene (DMBA) induced genotoxicity in Swiss albino mice. *Human & Experimental Toxicology*, *23*(11), 527–531.

Bhid, M. M., & Nitave, S. A. (2014). Roles of Emblica officinalis (Amla) in medicine. *World Journal of Pharmacy and Pharmaceutical Sciences (WJPPS)*, *3*(6), 604–615.

Brun, V., & Schumacher, T. (1994). *Traditional herbal medicine in Northern Thailand*. White Lotus.

Chatterjee, A., Chattopadhyay, S., & Bandyopadhyay, S. K. (2010). Biphasic effect of Phyllanthus emblica L. extract on NSAID-induced ulcer: an antioxidative trail weaved with immunomodulatory effect. *Evidence-Based Complementary and Alternative Medicine*, *2011* (1), 146808.

Chen, K. H., Lin, B. R., Chien, C. T., & Ho, C. H. (2011). Emblica officinalis Gaertn. attentuates N-nitrosodiethylamine-induced apoptosis, autophagy, and inflammation in rat livers. *Journal of Medicinal Food*, *14*(7–8), 746–755.

Chung, K. T., Wong, T. Y., Wei, C. I., Huang, Y. W., & Lin, Y. (1998). Tannins and human health: a review. *Critical Reviews in Food Science and Nutrition*, *38*(6), 421–464.

Colditz, G. A., Wolin, K. Y., & Gehlert, S. (2012). Applying what we know to accelerate cancer prevention. *Science Translational Medicine*, *4*(127), 127rv4.

Cragg, G. M., Grothaus, P. G., & Newman, D. J. (2009). Impact of natural products on developing new anti-cancer agents. *Chemical Reviews*, *109*(7), 3012–3043.

Cuzick, J. (2017 Aug). Preventive therapy for cancer. *Lancet Oncology*, *18*(8), e472–e482. doi: 10.1016/S1470–2045(17)30536-3.

Dasaroju, S., & Gottumukkala, K. M. (2014). Current trends in the research of Emblica officinalis (Amla): a pharmacological perspective. *International Journal of Pharmaceutical Sciences Review*, *24*(2), 150–159.

De, A., De, A., Papasian, C., Hentges, S., Banerjee, S., Haque, I., & Banerjee, S. K. (2013). Emblica officinalis extract induces autophagy and inhibits human ovarian cancer cell proliferation, angiogenesis, growth of mouse xenograft tumors. *PloS One*, *8*(8), e72748.

De, A., De, A., Sharma, R., Suo, W., & Sharma, M. (2020). Sensitization of carboplatinum-and taxol-resistant high-grade serous ovarian cancer cells carrying p53, BRCA1/2 mutations by Emblica officinalis (Amla) via multiple targets. *Journal of Cancer*, *11*(7), 1927.

De, A., Powers, B., De, A., Zhou, J., Sharma, S., Van Veldhuizen, P., Bansal, A., Sharma, R., & Sharma, M. (2016). Emblica officinalis extract downregulates pro-angiogenic molecules via upregulation of cellular and exosomal miR-375 in human ovarian cancer cells. *Oncotarget*, *7*(21), 31484–31500.

Deep, G., Dhiman, M., Rao, A. R., & Kale, R. K. (2005). Chemopreventive potential of Triphala (a composite Indian drug) on benzo (a) pyrene induced forestomach tumorigenesis in murine tumor model system. *Journal of Experimental & Clinical Cancer Research: CR*, *24*(4), 555–563.

Dhir, H., Roy, A. K., Sharma, A., & Talukder, G. (1990). Modification of clastogenicity of lead and aluminium in mouse bone marrow cells by dietary ingestion of Phyllanthus emblica fruit extract. *Mutation Research/Genetic Toxicology*, *241*(3), 305–312.

Diwan, G., Sinha, K., Lal, N., & Rangare, N. R. (2018). Tradition and medicinal value of Indian gooseberry: a review. *Journal of Pharmacognosy and Phytochemistry*, *7*(1), 2326–2333.

Edderkaoui, M., Odinokova, I., Ohno, I., Gukovsky, I., Go, V. L. W., Pandol, S. J., & Gukovskaya, A. S. (2008). Ellagic acid induces apoptosis through inhibition of nuclear factor κB in pancreatic cancer cells. *World Journal of Gastroenterology: WJG*, *14*(23), 3672.

Frei, B. (1994). Reactive oxygen species and antioxidant vitamins: mechanisms of action. *The American Journal of Medicine*, *97*(3), S5–S13.

Fridlender, M., Kapulnik, Y., & Koltai, H. (2015). Plant derived substances with anti-cancer activity: from folklore to practice. *Frontiers in Plant Science*, *6*, 799.

GuoX, Ni J, Liu X, Xue J, Wang X. (2013). Phyllanthus emblica L. fruit extract induces chromosomal instability and suppresses necrosis in human colon cancer cells. *International Journal for Vitamin and Nutrition Research*, *83*(5), 271–80. doi: 10.1024/0300–9831/a000169.

Guo, X., & Wang, X. (2016). Phyllanthus emblica fruit extract activates spindle assembly checkpoint, prevents mitotic aberrations and genomic instability in human colon epithelial NCM460 cells. *International Journal of Molecular Sciences*, *17*(9), 1437.

Haque, R., Bin-Hafeez, B., Ahmad, I., Parvez, S., Pandey, S., & Raisuddin, S. (2001). Protective effects of Emblica officinalis Gaertn. in cyclophosphamide-treated mice. *Human & Experimental Toxicology*, *20*(12), 643–650.

Hari Kumar, K. B., Sabu, M. C., Lima, P. S., & Kuttan, R. (2004). Modulation of hematopoietic system and antioxidant enzymes by Emblica officinalis Gaertn and its protective role against γ-radiation induced damages in mice. *Journal of Radiation Research*, *45*(4), 549–555.

Hasan, M. R., Islam, M. N., & Islam, M. R. (2016). Phytochemistry, pharmacological activities and traditional uses of Emblica officinalis: a review. *International Current Pharmaceutical Journal*, *5*(2), 14–21.

Hegde, S. K., Rao, S., Rao, P., Raghu, S. V., Meera, S., & Baliga, M. S. (2022). Aqueous extract of Emblica officinalis Linn (Indian gooseberry) in combination with iodine is more efficacious than iodine alone in mitigating mucositis in head and neck cancer patients undergoing curative radiotherapy: retrospective observations. *Indian Journal of Otolaryngology and Head & Neck Surgery*, *74*(3), 6330–6338.

Ho, H. H., Chang, C. S., Ho, W. C., Liao, S. Y., Lin, W. L., & Wang, C. J. (2013). Gallic acid inhibits gastric cancer cells metastasis and invasive growth via increased expression of RhoB, downregulation of AKT/small GTPase signals and inhibition of NF-κB activity. *Toxicology and Applied Pharmacology*, *266*(1), 76–85.

Ismail, T., Calcabrini, C., Diaz, A. R., Fimognari, C., Turrini, E., Catanzaro, E.,... & Sestili, P. (2016). Ellagitannins in cancer chemoprevention and therapy. *Toxins*, *8*(5), 151.

Jagetia, G. C., Baliga, M. S., Malagi, K. J., & Kamath, M. S. (2002). The evaluation of the radioprotective effect of Triphala (an ayurvedic rejuvenating drug) in the mice exposed to γ-radiation. *Phytomedicine*, *9*(2), 99–108.

Jain, S. K., & Khurdiya, D. S. (2004). Vitamin C enrichment of fruit juice based ready-to-serve beverages through blending of Indian gooseberry (Emblica officinalis Gaertn.) juice. *Plant Foods for Human Nutrition*, *59*(2), 63–66.

Jeena, K. J., Joy, K. L., & Kuttan, R. (1999). Effect of Emblica officinalis, Phyllanthus amarus and Picrorrhiza kurroa on N-nitrosodiethylamine induced hepatocarcinogenesis. *Cancer Letters*, *136*(1), 11–16.

Jeevangi, S., Manjunath, S., & Sakhare, P. M. (2013). A study of anti-hyperlipidemia, hypolipedimic and anti-atherogenic activity of fruit of Emblica officinalis (amla) in high fat fed albino rats. *International Journal of Medical Research & Health Sciences*, *2*(1), 70–77.

Jena, G. B., Nemmani, K. V., Kaul, C. L., & Ramarao, P. (2003). Protective effect of a polyherbal formulation (Immu-21) against cyclophosphamide-induced mutagenicity in mice. *Phytotherapy Research: An International Journal Devoted to Pharmacological and Toxicological Evaluation of Natural Product Derivatives*, *17*(4), 306–310.

Jindal, A., Soyal, D., Sharma, A., & Goyal, P. K. (2009). Protective effect of an extract of Emblica officinalis against radiation-induced damage in mice. *Integrative Cancer Therapies*, *8*(1), 98–105.

Jose, J. K., Kuttan, G., & Kuttan, R. (2001). Antitumor activity of Emblica officinalis. *Journal of Ethnopharmacology*, *75*(2–3), 65–69.

Juranić, Z., & Žižak, Ž. (2005). Biological activities of berries: from antioxidant capacity to anti-cancer effects. *Biofactors*, *23*(4), 207–211.

Kaur, S., Michael, H., Arora, S., Härkönen, P. L., & Kumar, S. (2005). The in vitro cytotoxic and apoptotic activity of Triphala—an Indian herbal drug. *Journal of Ethnopharmacology*, *97*(1), 15–20.

Khan, K. H. (2009). Roles of Emblica officinalis in medicine - a review. *Botany Research International*, *2*(4), 218–228.

Khan, M. S., Qais, F. A., & Ahmad, I. (2019). Indian berries and their active compounds: therapeutic potential in cancer prevention. In *New look to phytomedicine* (pp. 179–201). Academic Press.

Khan, M. T., Lampronti, I., Martello, D., Bianchi, N., Jabbar, S., Choudhuri, M. S. K. & Gambari, R. (2002). Identification of pyrogallol as an antiproliferative compound present in extracts from the medicinal plant Emblica officinalis: effects on in vitro cell growth of human tumor cell lines. *International Journal of Oncology*, *21*(1), 187–192.

Kim, H. J., Kim, J., Kang, K. S., Lee, K. T., & Yang, H. O. (2014). Neuroprotective effect of chebulagic acid via autophagy induction in SH-SY5Y cells. *Biomolecules & Therapeutics*, *22*(4), 275.

Kudva, A. K., Raghu, S. V., Rao, S., Venkatesh, P., Hegde, S. K., Baliga-Rao, M. P.,... & Baliga, M. S. (2022). Indian indigenous fruits as radioprotective agents: past, present and future. *Anti-Cancer Agents in Medicinal Chemistry (Formerly Current Medicinal Chemistry-Anti-Cancer Agents)*, *22*(1), 53–63.

Kulkarni, K. V., & Ghurghure, S. M. (2018). Indian gooseberry (Emblica officinalis): complete pharmacognosy review. *International Journal of Chemistry Studies*, *2*(2), 5–11.

Liang, C. Z., Zhang, X., Li, H., Tao, Y. Q., Tao, L. J., Yang, Z. R.,... & Tao, H. M. (2012). Gallic acid induces the apoptosis of human osteosarcoma cells in vitro and in vivo via the regulation of mitogen-activated protein kinase pathways. *Cancer Biotherapy and Radiopharmaceuticals*, *27*(10), 701–710.

Lim, D. W., Kim, J. G., & Kim, Y. T. (2016). Analgesic effect of Indian gooseberry (Emblica officinalis fruit) extracts on postoperative and neuropathic pain in rats. *Nutrients*, *8*(12), 760.

López-Gómez, M., Malmierca, E., de Górgolas, M., & Casado, E. (2013). Cancer in developing countries: the next most preventable pandemic. The global problem of cancer. *Critical Reviews in Oncology/Hematology*, *88*(1), 117–122.

Madhuri, S., Pandey, G., & Verma, K. S. (2011). Antioxidant, immunomodulatory and anticancer activities of Emblica officinalis: an overview. *International Research Journal of Pharmacy*, *2*(8), 38–42.

Mahata, S., Pandey, A., Shukla, S., Tyagi, A., Husain, S. A., Das, B. C., & Bharti, A. C. (2013). Anticancer activity of Phyllanthus emblica Linn.(Indian gooseberry): inhibition of transcription factor AP-1 and HPV gene expression in cervical cancer cells. *Nutrition and Cancer*, *65*(sup1), 88–97.

Makena, P. S., & Chung, K. T. (2007). Effects of various plant polyphenols on bladder carcinogen benzidine-induced mutagenicity. *Food and Chemical Toxicology*, *45*(10), 1899–1909.

Mir, M. A., Hamdani, S. S., Sheikh, B. A., & Mehraj, U. (2019). Recent advances in metabolites from medicinal plants in cancer prevention and treatment. *Current Immunology Reviews*, *15*(2), 185–201.

Mirunalini, S., & Krishnaveni, M. (2010). Therapeutic potential of Phyllanthus emblica (amla): the ayurvedic wonder. *Journal of Basic and Clinical Physiology and Pharmacology*, *21*(1), 93–105.

Naiki -Ito, A., Chewonarin, T., Tang, M., Pitchakarn, P., Kuno, T., Ogawa, K., & Takahashi, S. (2015). Ellagic acid, a component of pomegranate fruit juice, suppresses androgen-dependent prostate carcinogenesis via induction of apoptosis. *The Prostate*, *75*(2), 151–160.

Ngamkitidechakul, C., Jaijoy, K., Hansakul, P., Soonthornchareonnon, N., & Sireeratawong, S. (2010). Antitumor effects of Phyllanthus emblica L.: induction of cancer cell apoptosis and inhibition of in vivo tumor promotion and in vitro invasion of human cancer cells. *Phytotherapy Research*, *24*(9), 1405–1413.

NHB. (2018). *National Horticulture Board production statistics of 2018*. https://nhb.gov.in/Statistics.aspx?enc=WkegdyuHokljEtehnJoq0KWLU79sOQCy?W4MfOk01GFOWQSEvtp9tNHHoiv3p49g. Accessed 17 December 2019.

Nosal'ova, G., Mokrý, J., & Hassan, K. T. (2003). Antitussive activity of the fruit extract of Emblica officinalis Gaertn. (Euphorbiaceae). *Phytomedicine*, *10*(6–7), 583–589.

Omran, A. R. (1971). The epidemiological transition: a theory of the epidemiology of population change. *Millbank Memorial Fund Quarterly*, *49*, 509–538.

Ortega, A. M. M., & Campos, M. R. S. (2019). Medicinal plants and their bioactive metabolites in cancer prevention and treatment. In Health Benefits and Potential Applications. *Bioactive compounds* (pp. 85–109). Woodhead Publishing.

Pareek, S., Shikov, A. N., Pozharitskaya, O. N., Makarov, V. G., González-Aguilar, G. A., Ramalho, S. A., & Narain, N. (2017). Indian gooseberry (Emblica officinalis gaertn.). *Fruit and Vegetable Phytochemicals: Chemistry and Human Health, 2nd Edition*, 1077–1106.

Penolazzi, L., Lampronti, I., Borgatti, M., Khan, M. T. H., Zennaro, M., Piva, R., & Gambari, R. (2008). Induction of apoptosis of human primary osteoclasts treated with extracts from the medicinal plant Emblica officinalis. *BMC Complementary and Alternative Medicine*, *8*(1), 1–11.

Pinmai, K., Chunlaratthanabhorn, S., Ngamkitidechakul, C., Soonthornchareon, N., & Hahnvajanawong, C. (2008). Synergistic growth inhibitory effects of Phyllanthus emblica and Terminalia bellerica extracts with conventional cytotoxic agents: doxorubicin and cisplatin against human hepatocellular carcinoma and lung cancer cells. *World Journal of Gastroenterology*, *14*, 1491–1497.

Poltanov, E. A., Shikov, A. N., Dorman, H. D., Pozharitskaya, O. N., Makarov, V. G., Tikhonov, V. P., & Hiltunen, R. (2009). Chemical and antioxidant evaluation of Indian gooseberry (Emblica officinalis Gaertn., syn. Phyllanthus emblica L.) supplements. *Phytotherapy Research: An International Journal Devoted to Pharmacological and Toxicological Evaluation of Natural Product Derivatives*, *23*(9), 1309–1315.

Priya, F. F., & Islam, M. S. (2019). Phyllanthus emblica Linn.(Amla)—a natural gift to humans: an overview. *Journal of Diseases and Medicinal Plants*, *5*, 1–9.

Rajak, S., Banerjee, S. K., Sood, S., Dinda, A. K., Gupta, Y. K., Gupta, S. K., & Maulik, S. K. (2004). Emblica officinalis causes myocardial adaptation and protects against oxidative stress in ischemic-reperfusion injury in rats. *Phytotherapy Research: An International Journal Devoted to Pharmacological and Toxicological Evaluation of Natural Product Derivatives*, *18*(1), 54–60.

Rajeshkumar, N. V., Pillai, M. R., & Kuttan, R. (2003). Induction of apoptosis in mouse and human carcinoma cell lines by Emblica officinalis polyphenols and its effect on chemical carcinogenesis. *Journal of Experimental & Clinical Cancer Research: CR*, *22*(2), 201–212.

Rosarin, F. S., Arulmozhi, V., Nagarajan, S., & Mirunalini, S. (2013). Antiproliferative effect of silver nanoparticles synthesized using amla on Hep2 cell line. *Asian Pacific Journal of Tropical Medicine*, *6*(1), 1–10.

Sancheti, G., Jindal, A., Kumari, R., & Goyal, P. K. (2005). Chemopreventive action of Emblica officinalis on skin carcinogenesis in mice. *Asian Pacific Journal of Cancer Prevention: APJCP*, *6*(2), 197–201.

Sandhya, T., Lathika, K. M., Pandey, B. N., & Mishra, K. P. (2006). Potential of traditional ayurvedic formulation, Triphala, as a novel anticancer drug. *Cancer Letters*, *231*(2), 206–214.

Scartezzini, P. (2000). Review on some plants of Indian traditional medicine with antioxidant activity. *Journal of Ethnopharmacology*, *71*(1–2), 23–43.

Scartezzini, P., Antognoni, F., Raggi, M. A., Poli, F., & Sabbioni, C. (2006). Vitamin C content and antioxidant activity of the fruit and of the ayurvedic preparation of Emblica officinalis Gaertn. *Journal of Ethnopharmacology*, *104*(1–2), 113–118.

Seyfried, T. N., & Shelton, L. M. (2010). Cancer as a metabolic disease. *Nutrition & Metabolism*, *7*(1), 1–22.

Sharma, N., Trikha, P., Athar, M., & Raisuddin, S. (2000a). In vitro inhibition of carcinogen-induced mutagenicity by Cassia occidentalis and Emblica officinalis. *Drug and Chemical Toxicology*, *23*(3), 477–484.

Sharma, N., Trikha, P., Athar, M., & Raisuddin, S. (2000b). Inhibitory effect of Emblica officinals on the in vivo clastogenicity of benzo alpyrene and acyclophosphamide in mice. *Human & Experimental Toxicology*, *19*(6), 377–384.

Shivananjappa, M. M., & Joshi, M. K. (2012). Influence of Emblica officinalis aqueous extract on growth and antioxidant defense system of human hepatoma cell line (HepG2). *Pharmaceutical Biology*, *50*(4), 497–505.

Sidhu, S., Pandhi, P., Malhotra, S., Vaiphei, K., & Khanduja, K. L. (2011). Beneficial effects of Emblica officinalis in L-arginine-induced acute pancreatitis in rats. *Journal of Medicinal Food*, *14*(1–2), 147–155

Siegel, R. L., Miller, K. D., Fuchs, H. E., & Jemal, A. (2022). Cancer statistics, 2022. *CA: A Cancer Journal for Clinicians*. *72*(1), 7–33.

Singh, E., Sharma, S., Pareek, A., Dwivedi, J., Yadav, S., & Sharma, S. (2012). Phytochemistry, traditional uses and cancer chemopreventive activity of Amla (Phyllanthus emblica): the Sustainer. *Journal of Applied Pharmaceutical Science*, *2*, 176–183.

Singh, I., Sharma, A., Nunia, V., & Goyal, P. K. (2005). Radioprotection of Swiss albino mice by Emblica officinalis. *Phytotherapy Research: An International Journal Devoted to Pharmacological and Toxicological Evaluation of Natural Product Derivatives*, *19*(5), 444–446.

Singh, S., Singh, A. K., & Joshi, H. K. (2006). Standardization of maturity indices in Indian gooseberry (Emblica afficitlalis) under semi-arid conditions of Gujarat. *The Indian Journal of Agricultural Sciences*, *76*(10), 591–595.

Smith, G. L., Lopez-Olivo, M. A., Advani, P. G., Ning, M. S., Geng, Y., Giordano, S. H., & Volk, R. J. (2019). Financial burdens of cancer treatment: a systematic review of risk factors and outcomes. *Journal of the National Comprehensive Cancer Network JNCCN*, *17*(10), 1184–1192. doi: 10.6004/jnccn.2019.7305

Sosa, V., Moliné, T., Somoza, R., Paciucci, R., Kondoh, H., & LLeonart, M. E. (2013). Oxidative stress and cancer: an overview. *Ageing Research Reviews*, *12*(1), 376–390.

Srivasuki, K. P. (2012). Nutritional and health care benefits of Amla. *Journal of Pharmacognosy*, *3*(2), 147–151.

Sudhakar , A. (2009). History of cancer, ancient and modern treatment methods. *Journal of Cancer Science & Therapy*, *1*(2), 1.

Sultana, S., Ahmed, S., & Jahangir, T. (2008). Emblica officinalis and hepatocarcinogenesis: a chemopreventive study in Wistar rats. *Journal of Ethnopharmacology*, *118*(1), 1–6.

Sung, H., Ferlay, J., Siegel, R. L., Laversanne, M., Soerjomataram, I., Jemal, A., & Bray, F. (2021). Global cancer statistics 2020: GLOBOCAN estimates of incidence and mortality worldwide for 36 cancers in 185 countries. *CA: A Cancer Journal for Clinicians*, *71*(3), 209–249.

Tarapore, R. S., Siddiqui, I. A., & Mukhtar, H. (2012). Modulation of Wnt/β-catenin signaling pathway by bioactive food components. *Carcinogenesis*, *33*(3), 483–491.

Thakur, R. S., Puri, H. S., & Husain, A. (1989). *Major medicinal plants of India*. Central Institute of Medicinal and Aromatic Plants, Lucknow, India.

Vadde, R., Radhakrishnan, S., Kurundu, H. E. K., Reddivari, L., & Vanamala, J. K. (2016). Indian gooseberry (Emblica officinalis Gaertn.) suppresses cell proliferation and induces apoptosis in human colon cancer stem cells independent of p53 status via suppression of c-Myc and cyclin D1. *Journal of Functional Foods*, *25*, 267–278.

Vadhanam, M. V., Ravoori, S., Aqil, F., & Gupta, R. C. (2011). Chemoprevention of mammary carcinogenesis by sustained systemic delivery of ellagic acid. *European Journal of Cancer Prevention*, *20*(6), 484–491.

Variya, B. C., Bakrania, A. K., & Patel, S. S. (2016). Emblica officinalis (Amla): a review for its phytochemistry, ethnomedicinal uses and medicinal potentials with respect to molecular mechanisms. *Pharmacological Research, 111*, 180–200.

Vasant, B. S., Bhaskarrao, D. A., & Bhanudas, S. R. (2013). Emblica officinalis-the wonder of ayurvedic medicine. *World Journal of Pharmaceutical Sciences, 3*(1), 285–306.

Vasant, R. A., & Narasimhacharya, A. V. R. L. (2012). Amla as an antihyperglycemic and hepato-renal protective agent in fluoride induced toxicity. *Journal of Pharmacy & Bioallied Sciences, 4*(3), 250.

Wattanapitayakul , S. K., Chularojmontri, L., Herunsalee, A., Charuchongkolwongse, S., Niumsakul, S., & Bauer, J. A. (2005). Screening of antioxidants from medicinal plants for cardioprotective effect against doxorubicin toxicity. *Basic & Clinical Pharmacology & Toxicology, 96*(1), 80–87.

Wu, S., Powers, S., Zhu, W., & Hannun, Y. A. (2016). Substantial contribution of extrinsic risk factors to cancer development. *Nature, 529*(7584), 43–47.

Yadav, A., Sing, U., & Yadav, A. (2017a). 2. Aonla, physico-chemical and microbial quality during storage - a review by Anuj Yadav, Upendra Singh, Neha Nadga and Ashwani Yadav. *International Research Journal of Chemistry, 18*, 14.

Yadav, S. S., Singh, M. K., Singh, P. K., & Kumar, V. (2017b). Traditional knowledge to clinical trials: a review on therapeutic actions of Emblica officinalis. *Biomedicine & Pharmacotherapy, 93*, 1292–1302.

Yahayo, W., Supabphol, A., & Supabphol, R. (2013). Suppression of human fibrosarcoma cell metastasis by Phyllanthus emblica extract in vitro. *Asian Pacific Journal of Cancer Prevention, 14*(11), 6863–6867.

Yahia, E. M. (Ed.). (2011). *Postharvest biology and technology of tropical and subtropical fruits: fundamental issues*. Elsevier.

Yang, C. J., Wang, C. S., Hung, J. Y., Huang, H. W., Chia, Y. C., Wang, P. H.,... & Huang, M. S. (2009). Pyrogallol induces G2-M arrest in human lung cancer cells and inhibits tumor growth in an animal model. *Lung Cancer, 66*(2), 162–168.

Yıldırım, I., & Kutlu, T. (2015). Anticancer agents: saponin and tannin. *International Journal of Biological and Chemical Sciences, 9*, 332–340.

Yokozawa, T., Kim, H. Y., Kim, H. J., Tanaka, T., Sugino, H., Okubo, T. & Juneja, L. R. (2007). Amla (Emblica officinalis Gaertn.) attenuates age-related renal dysfunction by oxidative stress. *Journal of Agricultural and Food Chemistry, 55*(19), 7744–7752.

Zhang, Y. B., Pan, X. F., Chen, J., Cao, A., Zhang, Y. G., Xia, L., & Pan, A. (2020). Combined lifestyle factors, incident cancer, and cancer mortality: a systematic review and meta-analysis of prospective cohort studies. *British Journal of Cancer, 122*(7), 1085–1093

Zhang, Y. J., Nagao, T., Tanaka, T., Yang, C. R., Okabe, H., & Kouno, I. (2004). Antiproliferative activity of the main constituents from Phyllanthus emblica. *Biological and Pharmaceutical Bulletin, 27*(2), 251–255.

Zhao, T., Sun, Q., Marques, M., & Witcher, M. (2015). Anticancer properties of Phyllanthus emblica (Indian gooseberry). *Oxidative Medicine and Cellular Longevity, 2015, 2015*(1), 950890.

Zhu, X., Wang, J., Ou, Y., Han, W., & Li, H. (2013). Polyphenol extract of Phyllanthus emblica (PEEP) induces inhibition of cell proliferation and triggers apoptosis in cervical cancer cells. *European Journal of Medical Research, 18*(1), 1–5.

8 Quinoa and Sea Buckthorn
Underutilized Crops with Diverse Potential

Geetanjali Sageena
Indian Council of Medical Research, New Delhi, India.

Khushboo Sharma
Banaras Hindu University, Varanasi, UP

Manish Sharma
University of Allahabad, Allahabad, India

8.1 INTRODUCTION

Pseudocereals are so named because they resemble cereals physically and because their edible seeds have a high starch content that can be ground into flour (Alvarez-Jubete et al. 2010). Utilizing these grains to create wholesome and nutrient-dense foods has compelled food manufacturers to create innovative, effective processing methods. Pseudocereals are also featured in the "International AACC list of recognised grains" (Gordon 2006), with quinoa (*Chenopodium quinoa*) being the most significant species, followed by buckwheat (*Fagopyrum esculentum*) and amaranth (*Amaranthus* sp.). Contrary to true cereals (Poaceae family), which are monocotyledonous species, pseudocereals are dicotyledonous species. These pseudocereals have drawn a lot of attention recently as good candidates for gluten-free products due to their well-known high nutritional value (quantity and quality of proteins they contain) and potential health advantages (Das 2016; Shewry 2002).

Since ancient times, underutilized wild plants have served as significant sources of nutrition and medicine, enhancing diet diversity and reducing malnutrition. These plants have been used to cure a variety of illnesses in traditional Chinese and Indian medicine, including Ayurvedic, Unani, and Siddha. Investigations into the phytochemistry of these plants have shown that they contain vital nutrients and bioactive substances that are responsible for their pharmacological actions. Researchers from all over the world are interested in pseudocereal plants because of their potential health benefits, nutritional worth, and socioeconomic importance.

DOI: 10.1201/9781003477792-8

Among pseudocereals, quinoa is promising for human consumption and nutrition because of its grains' high protein content; quantity of vital amino acids; and variety of vitamins, minerals, and saponins. Sea buckthorn (SBT) is found growing extensively in the trans-Himalayan regions of Ladakh (UT) and Lahaul-Spiti in Himachal Pradesh, and its biodiversity is also abundant in the Himalayan slopes of Uttarakhand. In addition, Sikkim and Arunachal Pradesh have some areas where it grows. Additionally, adding these plants to staple foods can boost their micronutrient and bioactive content. The discovery of these substances via a metabolomics technique might lead to the creation of innovative medications that are more economical and safer. The current chapter highlights the potential therapeutic role of quinoa and SBT species in India.

8.2 BACKGROUND OF QUINOA

Quinoa (*C. quinoa*), termed a pseudocereal, is an annual herbaceous plant that originally grew on the Pacific slopes of the Andes in South America but is now considered a member of the *Amaranthaceae* family and the goosefoot genus (Ramesh 2017; Kadereit et al. 2003). There are more than 250 species of *Chenopodium* seen worldwide (Vega et al. 2010). According to Food and Agriculture Organization Corporate Statistical Database (2013), Bolivia in South America produces 46% of the world's quinoa, followed by Peru with 42% and the United States of America with 6.3%. According to Srinivasa Rao (2015), quinoa was grown in India on 440 hectares of land with an annual yield of 1,053 tonnes. The plant can thrive at high altitudes in mountainous regions where maize cannot because it can withstand harsh environmental conditions such as salinity, cold, sun radiation, and drought. Due to this crop's high adaptation to different agro-climatic habitats and edaphic conditions, quinoa production has increased globally in recent years in the USA, Europe, Canada, China, and India (González et al. 2012; Jacobsen 2011).

8.2.1 History

When quinoa was initially cultivated around 5,000 years ago, the Incas called it "the mother of all grains." It was considered a gift from their gods and used even for therapeutic purposes. For the Incas and their indigenous descendants, the Quechua and Aymara, it remained a vital source of food. According to mythology, the Inca emperor ceremoniously sowed the first quinoa seedlings each year. Earlier, quinoa seeds were used as a grain, an ingredient in soups, and even fermented to produce beer or chichi, an Andean traditional beverage (Vega et al. 2010; Cooper 2015). As previously indicated, quinoa is farmed or tested in 95 different nations throughout the world (Bazile, Didier, 2015). Quinoa is a grain that has long been popular as a health food in North America, Europe, Australia, Japan, South America's Andean Highlands, and India (http://whfoods.org//genpage.php?tname=foodspice&dbid=142).

8.2.2 Nutritional Components

The quinoa seeds have an embryo that is rich in protein and oil, surrounding a core perisperm where the carbohydrate reserves are concentrated in the seed coat and endosperm.

A bitter chemical called saponins is present in the fruit's pericarp and can be eliminated by washing or mechanical abrasion. Other names for this de-saponification procedure include milling (Kumpun et al. 2011), pearling (Gómez-Caravaca et al. 2014), and dehusking (Miranda et al. 2012). It belongs to the "whole grains" category because of the nutritional content of its ingredients (McKeown et al. 2013). One of the key benefits of quinoa is that, unlike other cereal grains, which are processed to remove the nutrient-rich germ and bran, quinoa is de-saponified to preserve the nutrient-rich embryo and to maintain the entire endosperm (Graf et al. 2015).

The quinoa grain is the only vegetable food that, according to the United Nations Organization for Agriculture and Food, offers all of the amino acids necessary for human life in sufficient amounts and is similar to milk. The range of the protein content is 7.47%–22.08%. The necessary fatty acids are abundant, and the oil content ranges from 1.8% to 9.5%. More than 80% of the quinoa protein can be digested. Quinoa also has natural antioxidants, including α -tocopherol (5.3 mg) and γ-tocopherol (2.6 mg) per 100 g of seed, as well as phytoestrogens that guard against heart disease, osteoporosis, breast cancer, and other women's issues brought on by menopause's loss of oestrogen (Bhargava et al. 2006). Similar to quinoa grains, quinoa greens (leaves, microgreens, and sprouts) are rich in nutrients and possess a number of health-promoting qualities, including anti-microbial, anti-cancer, anti-diabetic, anti-obesity, antioxidant, and cardio-beneficial properties (Hosny et al. 2022; Valenzuela Zamudio et al. 2022; Lápez-Cervantes et al. 2021; Park et al. 2017; Gawlik-Dziki et al. 2013; Paśko et al. 2009; Paśko et al. 2008). To make up for the dearth of animal proteins in their diet, the Incas used quinoa leaves and grains (Camaggio et al. 2014).

Being rich in many essential elements, it is regarded as one of the most well-liked health foods worldwide. Quinoa leaves have long been a favourite food of Andean inhabitants. Fresh leaves and delicate shoots can be either raw or cooked, together with quinoa sprouts in a salad (Suomela et al. 2022; Tang et al. 2015a, 2015b; Bhargava et al. 2006). Recent studies have shown that quinoa leaves have low levels of carbohydrates and high levels of protein, including all essential amino acids for humans. Additionally, quinoa leaves have high concentrations of potassium, manganese, and copper, as well as average concentrations of calcium, phosphorus, salt, and zinc (Villacrés et al. 2022; Adamczewska-Sowińska et al. 2021; Pathan et al. 2019; Ogungbenle 2003).

Quinoa was deemed by the Food Association Organisation to have a high nutritional value, a wide range of biological diversity, and the potential to significantly contribute to global food security (Bastidas et al. 2016). The nutritional and bioactive properties of quinoa grains have been the subject of numerous studies and review papers (Pathan et al. 2022; Melini and Melini 2021; Mohamed et al. 2021; El Hazzam et al. 2020). As per the -United States Department of Agriculture (2013), the nutritional quality of quinoa (uncooked) is detailed in Table 8.1. Quinoa benefits various high-risk consumers, from the elderly to kids; lactose-intolerant persons; and those suffering from diabetes, obesity, dyslipidemia, anaemia, and celiac disease. In addition, it has high nutritional value and gluten-free status (Ibrahem et al. 2021; Cao et al. 2020; Singh et al. 2016; Zevallos et al. 2014). These effects have been linked to quinoa's high protein, fibre, vitamin and mineral content, fatty acid composition,

TABLE 8.1
Nutritional Components of Uncooked Quinoa (USDA 2018)

Nutrient	Quantity	Unit (per 100g)
Water	71.6	g
Energy (kcal)	120	kcal
Proteins	4.4	g
Total fat	1.92	g
Carbohydrate, by difference	21.3	g
Fibre, total dietary	2.8	g
Sugars, total including NLEA	0.87	g
Starch	17.6	g
Minerals	**Quantity**	**Unit (per 100g)**
Calcium	17	mg
Iron	1.49	mg
Magnesium	64	mg
Phosphorus	152	mg
Potassium	172	mg
Sodium	7	mg
Zinc	1.09	mg
Copper	0.192	mg
Manganese	0.631	mg
Selenium	2.8	µg
Vitamins	**Quantity**	**Unit (per 100g)**
Vitamin B1 (thiamine)	0.107	mg
Vitamin B2 (riboflavin)	0.11	mg
Vitamin B3 (niacin)	0.412	mg
Vitamin B6 (pyridoxine)	0.123	mg
Folate total	42	µg
Folate food	42	µg
Folate DFE	42	µg
Choline, total	23	µg
Carotene, beta	3	µg
Vitamin A, IU	5	IU
Lutein + zeaxanthin	53	µg
Vitamin E (alpha-tocopherol)	0.63	mg
Tocopherol, beta	0.03	mg
Tocopherol, gamma	1.19	mg
Tocopherol, delta	0.11	mg
Lipids	**Quantity**	**Unit (per 100g)**
Fatty acids, total saturated	0.231	g
SFA 16:0	0.196	g
SFA 18:0	0.012	g
Fatty acids, total monounsaturated	0.528	g

(Continued)

TABLE 8.1 (*Continued*)
Nutritional Components of Uncooked Quinoa (USDA 2018)

Nutrient	Quantity	Unit (per 100 g)
MUFA 18:1	0.465	g
MUFA 20:1	0.031	g
MUFA 22:1	0.027	g
Fatty acids, total polyunsaturated	1.08	g
PUFA 18:2	0.974	g
PUFA 18:3	0.085	g
PUFA 20:4	0.003	g
PUFA 22:6 n-3 (DHA)	0.015	g
Essential Amino Acids	**Quantity**	**Unit (per 100 g)**
Tryptophan	0.052	g
Threonine	0.131	g
Isoleucine	0.157	g
Leucine	0.261	g
Lysine	0.239	g
Methionine	0.096	g
Cysteine	0.063	g
Phenylalanine	0.185	g
Tyrosine	0.083	g
Valine	0.185	g
Arginine	0.34	g
Histidine	0.127	g
Alanine	0.183	g
Aspartic acid	0.353	g
Glutamic acid	0.58	g
Glycine	0.216	g
Proline	0.24	g
Serine	0.176	g

SFA: saturated fatty acid; MUFA: monounsaturated fatty acid; PUFA: polyunsaturated fatty acid.

and especially the inclusion of several phytochemicals, which give quinoa a notable nutritional and health edge over other grains (Hussain et al. 2021; Navruz-Varli et al. 2016). The impacts of quinoa's bioactive components and the supporting data regarding their influence on human health will be outlined below.

8.2.3 Bioactive Compounds Found in Quinoa

8.2.3.1 Phenolic Compounds

A broad and varied class of substances known as phenols are made up of hydroxyl group(s) connected to at least one aromatic hydrocarbon ring attached. Phenolics have great antioxidant potential because of their high structural stability. The various classes of phenolic chemicals include phenolic acids, flavonoids, lignans, coumarins, phenols, phenylpropanoids, quinines, stilbenoids, and xanthones in general. Because of their excellent structural stability, phenolic compounds have a strong antioxidant potential (Woldemichael and Wink 2001). The quantity per 100 g of the dry weight of quinoa grains contains free phenolic compounds ranging from 167.2 to 308.3 mg gallic acid equivalents (Han et al. 2019; Harborne et al. 2000).

Flavonoids: The quinoa plant contains a wide range of flavonoids (Figure 8.1), such as protocatechuic acid, benzoic acid, ferulic acid, vanillic acid, p-coumaric acid, and syringic acid (HernándezLedesma 2019; Tsao 2010). Humans' health has been significantly improved by plant-based isoflavones. Lutz et al. (2013) validated the presence of daidzein and genistein in the range of 0.70–2.05 and 0.05–0.41 mg/100 g, respectively, among seeds of ten different types of quinoa. Isoflavones were first discovered in quinoa plants.

Phenolic Acids: Seven types of quinoa were found to have a free proportion of total phenolic contents that ranged from 53.5% to 78.0%, with ferulic acid and gallic acid being two of the more notable components. The seeds include derivatives of the flavonoids quercetin, rutin, and kaempferol (Paśko et al. 2009). Orientin (1.08 mg/g

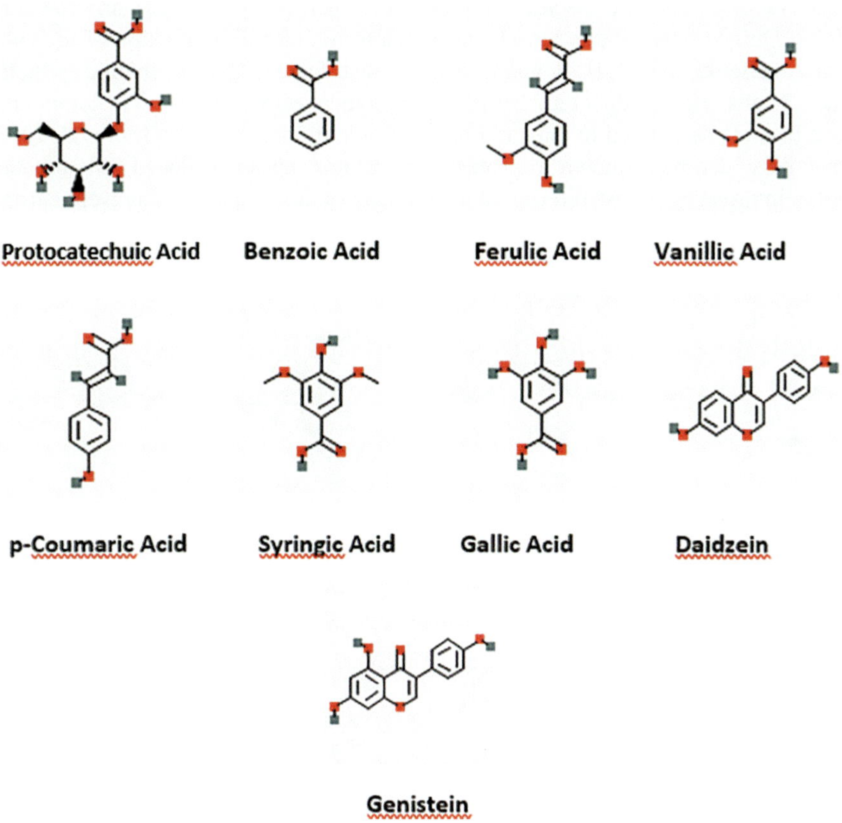

FIGURE 8.1 Some common flavonoids in quinoa.

dry weight), vitexin (0.71 mg/g dry weight), rutin (0.36 mg/g dry weight), as well as morin (88.9 g/g dry weight), hesperidin (1.86 g/g dry weight), and neohesperidin (1.93 g/g dry weight), were found to be present in quinoa seeds (Gorinstein et al. 2008), as shown in Figure 8.2. The concentrations of certain phenolic acids, such as ferulic, caffeic, and p-coumaric acids, are 251.5, 6.31, and 1.1 g/g dry weight, respectively (Tang et al. 2015a, b).

The quinoa grains also include bonded phenolics, which are so-called because they are linked to rod-shaped structural proteins as well as cell wall components like pectin, lignin, cellulose, and hemicellulose (arabinoxylans) (Samtiya et al. 2021; Abderrahim et al. 2015; Gómez et al. 2011; Da-Silva et al. 2007). In comparison to free phenolics, quinoa grains have a lower concentration of bound phenolics (Tang et al. 2015a, b; Renard et al. 1999). The majority of quinoa's bound phenolic chemicals are found in the leaves, not the seeds.

Phenolic molecules have strong antioxidant characteristics that can act in a variety of ways because of their stable chemical structure. They are cofactors of enzymes that catalyze oxidative reactions and act as reducers, free radical scavengers, and metal ion chelators; stabilize free radicals; stop radical chain reactions; inhibit oxidases, or other responses (Dini et al. 2010). Different types of quinoa include phenolic chemicals that are essential in reducing diabetes and obesity (Harborne et al. 2000). Quinoa polyphenols have been shown to inhibit the production of the cytokines interleukin (IL)-1, IL-8, and tumour necrosis factor (TNF) in colonic epithelial Caco-2 cells (Hemalatha et al. 2016). The biological effects of phenolic compounds have been demonstrated to play an important role in therapeutics (Graf et al. 2015). They have also been shown to protect mice against obesity-induced inflammation and to improve their gastrointestinal health. Due to their capacity to prevent chronic

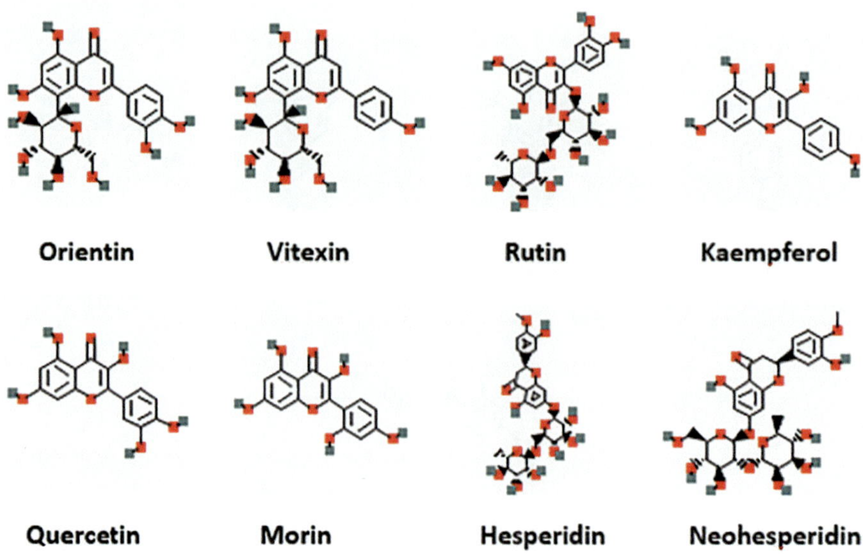

FIGURE 8.2 Phenolic acids present in quinoa.

diseases and their positive effects on health, phenolic compounds have attracted significant attention in research during the past 20 years (Han et al. 2019).

8.2.3.2 Phytoecdysteroids

Chemicals called phytoecdysteroids are used by plants as a defence mechanism. They are ecdysone analogues, an insect-moulting hormone. Over 500 phytoecdysteroid structures have been identified as of now (Moreau et al. 2018). According to Graf et al. (2015), at least 13 phytoecdysteroids have been found in quinoa seeds, some of which are shown in Figure 8.3.

The only pseudocereal with a considerable concentration of phytoecdysteroids, ranging from 138 to 570 g/g, is quinoa (Kumpun et al. 2011), which includes roughly 36 different types of phytoecdysteroids (Graf et al. 2016). According to research (Anthony et al. 2015; HernándezLedesma 2019), the most prevalent phytoecdysteroid makes up 62%–90% of all quinoa phytoecdysteroids and offers a variety of health-promoting properties. Among all the sources currently known, quinoa has been reported to have the highest output of 20 hydroxyecdysone (Graf et al. 2014; Kumpun et al. 2011). Others include 24(28)-dehydromakisterone A, Makisterone A, 24-epi-makisterone A, and 25,27-dehydroinokosterone, as well as polypodine B, 5 β-hydroxy-24(28)-dehydromakisterone A, and 24,25-dehydroinokosterone. Phytoecdysteroids have been utilized as a secure and successful alternative to anabolic steroids (Nsimba et al. 2008). Additionally, these phytoecdysteroids have antioxidant potential due to their metal ion chelating ability and free radical-scavenging activity, which makes them useful for preventing skin ageing (Lin et al. 2019). These bioactive substances are also very beneficial for promoting the growth of skeletal muscles, which is crucial for improving physical performance (Báthori et al. 2008). It has also been claimed that quinoa phytoecdysteroids may help prevent diabetes by reducing oxidative degradation and improving blood glucose transportation (Foucault et al. 2014).

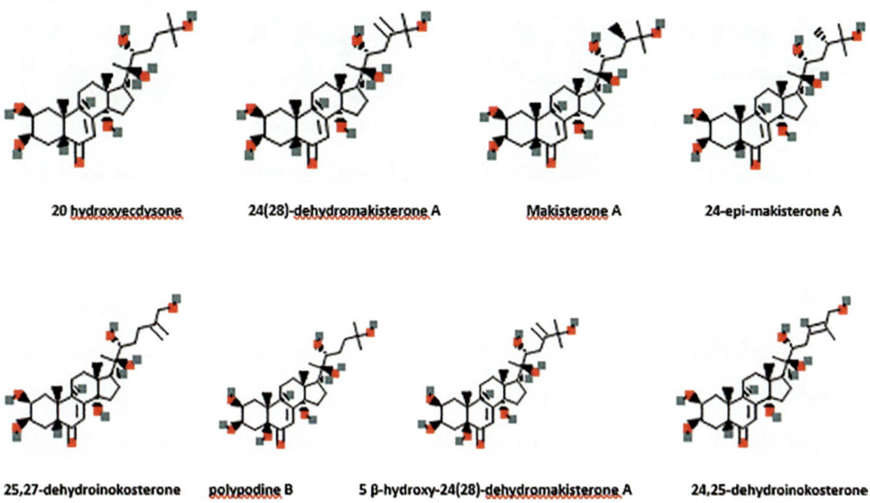

FIGURE 8.3 Phytoecdysteroids observed in quinoa.

8.3 SAPONINS

The outer layer (pericarp) of quinoa grains may contain saponins, secondary metabolites with a potent bitter taste (Alvarez-Jubete et al. 2010; El Hazzam et al. 2020). Chemically, saponins are made up of a steroidal aglycone or triterpenoid with one or more sugar moieties, such as glucose, galactose, arabinose, xylose, and glucuronic acid (Xia et al. 2014). Studies show that about 68 types of saponin compounds have been identified in crude quinoa seed extract using nano- High Performance Liquid Chromatography electrospray ionization multistage tandem mass spectrometry (Yendo et al. 2010). Saponins present in quinoa grains are highly diverse in their structural characteristics. According to Dini et al. 2010, quinoa grains can be classed as having "sweet" (0.11% saponins) or "bitter" (>0.11% saponins) saponin content. Hederagenin (HD), oleanolic acid (OA), phytolaccagenic acid (PA), serjanic acid, and 3-hydroxyolean-12-en-28-oic acid are the major sources of triterpene glycosides that make up quinoa saponins (HernándezLedesma 2019) (Figure 8.4). The three main aglycones are HD, PA, and OA saponins from quinoa (Lim et al. 2020).

8.4 BETANINS

The *Caryophyllales* order contains the nitrogenous plant pigments known as betalains. They can be split into yellow-orange or red-violet betacyanins or betaxanthins based on their chemical makeup, which together produce the orange and red colours

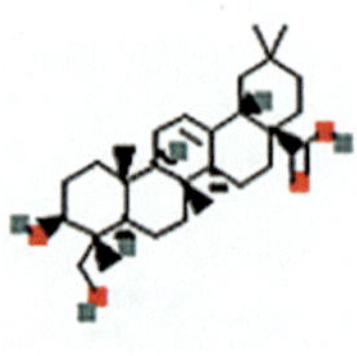
Hederagenin

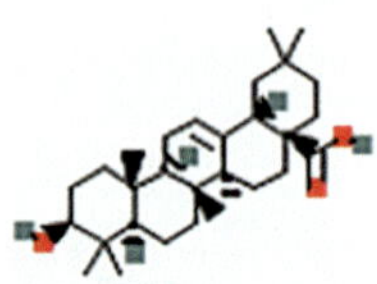
oleanolic acid

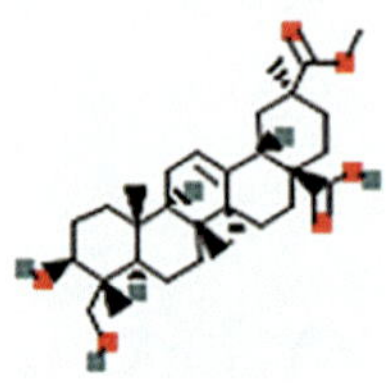
phytolaccagenic acid

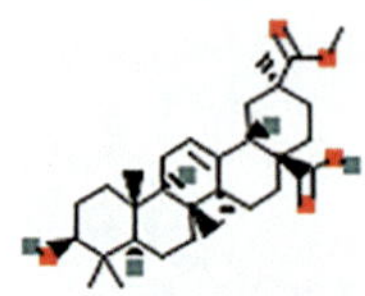
serjanic acid

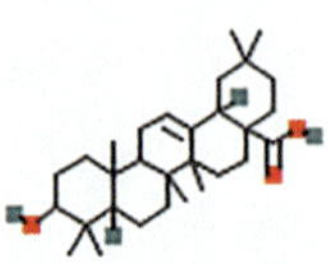
3-hydroxyolean-12-en-28-oic acid

FIGURE 8.4 Common saponins found in quinoa.

that coexist in nature. The European Union (additive E-162) and the U.S.Food and Drug Administration have approved betalain-containing extracts as natural colourants in food goods and pharmaceuticals (Gengatharan et al. 2015). New betacyanins and betaxanthins have been identified according to a recent study, which also linked their presence to the significant antioxidant and free radical-scavenging properties shown in grain extracts from 29 kinds of Peruvian quinoa (Aguilar-Tuesta et al. 2018). Additionally, betalains have grown in prominence as a component of functional foods in recent years due to their antioxidant, anti-cancer, anti-lipidemic, and anti-bacterial properties (Madadi et al. 2020, Gandía-Herrero et al. 2016, Repo-Carrasco-Valencia et al. 2010).

8.5 THERAPEUTIC VALUE

8.5.1 Clinical Trial Evidence

In a study by Navarro-Perez et al. (2017), it was examined how quinoa seeds affected blood triglyceride levels in obese and overweight people in a dose-dependent manner. The study assessed the lipid profiles, body compositions, and nutritional intake of quinoa seeds (25 or 50 grams per day) versus a control group for 12 weeks in adults. Participants who took 50 g of quinoa seeds per day showed a significant decline in serum triglycerides, but other indicators including total cholesterol, High Density Lipoproteins and Low Density Lipoproteins- remained stable. Higher serum triglycerides are frequently a risk factor for cardiovascular disease (CVD); therefore, the authors noticed a 36% reduction as a promising sign that quinoa consumption may help lower the risk of CVD. Another four-week randomized, cross-over trial investigated the impact of quinoa biscuits on CVD biomarkers. Researchers conducted a study on healthy older persons and discovered that the quinoa group had lower total and LDL cholesterol, body weight, and –body mass index in comparison to the control group (Pourshahidi et al. 2020). Despite the fact that the aforementioned research revealed modest alterations in biomarkers linked to CVD, the mechanism underlying such an effect is still not fully understood. In a meta-analysis study, it was observed by lowering adult blood triglycerides that quinoa supplementation at levels over 50 g may reduce the risk of CVD (Atefi et al. 2022). Table 8.2 lists the clinical studies that have been done so far to assess the health benefits using different biomarkers in human participants, demonstrating the health advantages of quinoa; nevertheless, the underlying mechanism for these benefits to health is mostly unknown.

8.5.2 Anti-Cancerous Potential of Quinoa

The polyphenolic chemicals saponin, betalain, carotenoids, and flavonoids, which are organic antioxidants with anti-cancer potential, are abundant in quinoa seeds and leaves (Pellegrini et al. 2017). Other research has discovered the anti-cancer effects of phytoecdysteroids, bioactive peptides, and quinoa polysaccharides (Hu et al. 2017). In a few investigations, quinoa was found to have anti-cancer effects. With an IC50 value of 14.6 g/mL, an *in vitro* study discovered that quinoa seeds have a promising cytotoxic impact against the liver cancer cell line HepG2 (Mohamed et al. 2019).

TABLE 8.2
Clinical Trial Evidence on Quinoa Consumption

Sr. No.	Key Biomarker Assessed	Reference
1	Lipid profile (total cholesterol, triglycerides, HDL, LDL) Glucose Hormones (leptin, adiponectin, insulin, C-peptide)	Navarro-Perez et al. (2017)
2	Lipid profile (total cholesterol, triglycerides, HDL, LDL) Glucose	Silva et al. (2018)
3	Lipid profile (total cholesterol, triglycerides, HDL) Serum antioxidants Serum PUFA	Pourshahidi et al. (2020)

The cytotoxic potential of pseudocereal seeds against the human cervical carcinoma cell lines C4-I, HTB-35, and HTB-34 was also examined in an *in vitro* investigation. A recent study opined that quinoa extract may be able to prevent colorectal cancer by lowering Wnt-mediated cell proliferation, which lowers the risk of the disease, decreasing angiogenesis through the PI3K/Akt pathway and inflammation through the NF-κB pathway (Borjac, Jamilah 2022) (Figure 8.5).

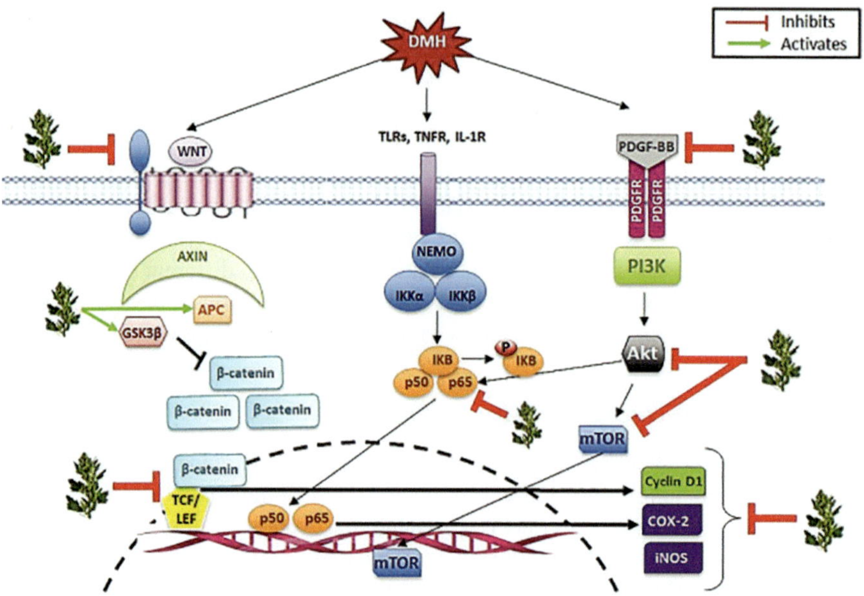

FIGURE 8.5 Effects of C. quinoa extract on various signalling pathways connected to colorectal cancer induced by 1,2 dimethylhydrazine (Borjac, Jamilah 2022).

8.5.3 Conclusion

One of the fascinating changes in the agro-food sector is the innovation of nutraceutical foods and their products. Researchers and scientists are under pressure to create food products that are high in protein content, gluten-free, and nutrient-dense as a result of people's increasing knowledge of nutrient-dense diets around the world. To sum up, quinoa is highly nutritious and contains a number of bioactive compounds, making it a viable ingredient in functional foods. Numerous studies have supported the quinoa diet's antioxidant, immunomodulatory, hypolipidemic, anti-cancer, hormone-regulating properties, and weight-regulating benefits, all of which support liver and cardiovascular health. Quinoa use has relatively little danger when weighed against its possible advantages. As a result, quinoa has a high potential to be marketed as a nutritious food. As a result, in recent years, interest has grown in the development of wholesome, nutritious goods made from pseudocereals for living a disease-free, healthy life. These products may also be helpful for celiac patients. As is well known, the main pseudocereals with a balanced nutritious composition of dietary protein, lipids, vitamins, minerals, dietary fibres, antioxidants, and bioactive components include amaranth, buckwheat, chia seeds, and quinoa.

8.6 BACKGROUND ABOUT SEA BUCKTHORN (SBT)

SBT is a plant that is widely used for nutraceutical and medicinal purposes (Ciesarová et al. 2020). SBT, previously known as *Elaeagnus rhamnoides* (L.) A. Nelson and/or *Hippophae rhamnoides* L., is a member of the *Elaeagnaceae* family. Geographically, it is distributed in the northwestern Himalayan region of India. It is primarily found in Himachal Pradesh's Lahaul-Spiti and Kinnaur, as well as Jammu and Kashmir's Ladakh and Kargil. At the global level, the major producer is Russia, followed by China, among other countries. It is also known as Siberian pineapple or seaberries. There are various bioactive compounds present in the bark, seeds, leaves, and berries. They are known to possess anti-inflammatory and anti-cancerous properties (Olsson 2004). It is used as traditional medicine widely in Asia (Suryakumar & Gupta 2011). The SBT has many bioactive compounds that are derived from berries, leaves, and bark. The berries are the source of juice, and the seeds are the source of oil (Table 8.3). The importance of this traditional plant lies in the fact that it is widely utilized due to its positive nutrient composition. It has been ascribed to be a rich source of antioxidants and immunomodulatory in nature (Olas et al. 2016, 2018; Olas 2016).

SBT extracts are broadly divided into phenolic and non-phenolic compounds. The phenolic compounds are further categorized as flavanols (isorhamnetin), tannins (proanthocyanidin), and phenolic acids (gallic acid). Non-phenolic compounds are split into lipophilic compounds (phytosterols, carotenoids), sugar derivatives, alcohols (ethyl-β-D-glucopyranoside, methyl-inositol), and organic acids (ascorbic acid) (Ma et al. 2022) (Figure 8.6).

TABLE 8.3
Characteristic Features of SBT

Characteristic (U.M.)	Range of Variation of the Values	Species/Variety
Fruit weight (mg)	270–480	*Indian Summer* subsp.
Moisture (%)	73.60–85.30	*Indian Summer* subsp.
	72.20–75.50	China var.
pH	2.50–2.73	*Carpatica* subsp.
Acidity (% malic acid)	1.64–1.74	*Carpatica* subsp.
Juice oil content (%)	0.26–1.43	China var.
	1.80–2.90 (pulp)	China var.
Unit weight	1.03–1.05	China var.

Source: Adapted from Gâtlan and Gutt (2021).

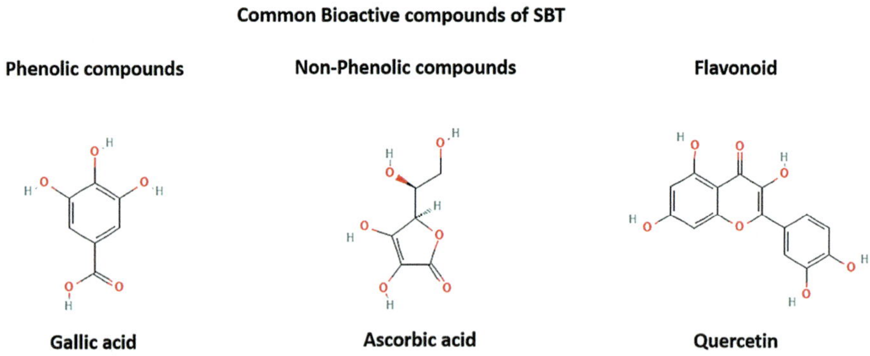

FIGURE 8.6 Common bioactive compounds extracted from SBT.

8.6.1 Nutrient Constituents and Bioactive Compounds

SBT, despite being a rich source of health-promoting nutrients with a diverse array of phytochemicals, is not a preferred source due to its sensory characteristics, which are strongly sour, bitter, and low in sugar content (Ma et al. 2022). However, the economic importance of this plant cannot be ignored. The SBT species, *H. rhamnoides* L., is an important source of natural antioxidants. Recently, HPLC analysis revealed that leaves and berries are rich in phenolic compounds, viz., quercetin and hydrocinnamic acid derivatives (Table 8.4). The other carotenoid compounds identified from berries were lutein, zeaxanthin, β-cryptoxanthin, *cis*-β-carotene, and β-carotene (Criste et al. 2020). Additionally, various subspecies of SBT, such as *H. rhamnoides* L., *H. rhamnoides* L. subsp. *sinensis* (Sinensis), *H. rhamnoides* L. subsp. *yunnanensis* (Yunnanensis), *H. rhamnoides* L. subsp. *mongolica* (Mongolica), and *H. rhamnoides* L. subsp. *turkestanica* (Turkestanica), have been evaluated for phytochemical analysis. The extracts from these subspecies were compared, and phenolic, antioxidant, and anti-proliferative properties were ascertained (Guo et al. 2017).

TABLE 8.4
Common Antioxidants of SBT

Part of the Plant	Antioxidant Compounds
Berries	Carotenoids, flavonoids, and organic acids
Seeds	Catechin, epicatechin, gallocatechin, and epigallocatechin
	Catechin(4α-8)catechin and catechin(4α-8)epicatechin
Oil	Carotenoids and β-sitosterol
	Monounsaturated fatty acids
	Tocopherols and tocotrienols
Leaves	Kaempferol-3-O-β-D-(6″-O-coumaryl)glycoside, 1-feruloyl-β-D-glucopyranoside, isorhamnetin-3-O-glucoside, quercet in-3-O-β-D-glucopyranoside, quercetin-3-O-β-Dglucopyranosyl-7-O-α-L-rhamnopyranoside and isorhamnetin-3-O-rutinoside

Source: Adapted from Gâtlan and Gutt (2021).

i. Phenolic compounds

One of the major phenolic compounds is flavonoids. They have an important role in preventing cancer. Flavonol is one of the abundant groups and is found in the glycosylated form of the aglycons isorhamnetin, quercetin, or myricetin. The sugar group determines the nature and quality of berries (Arimboor et al. 2012) (Figure 8.7).

ii. Tannins

They are categorized as hydrolysable and condensed tannins. For example, ellagitannins are a monomeric group of hydrolysable tannins, while the latter have more than one monomeric unit, like epigallocatechin units (Salminen & Karonen 2011).

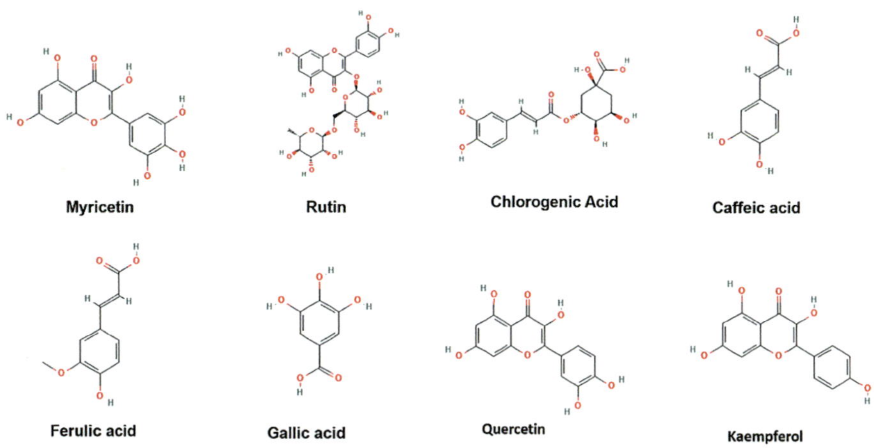

FIGURE 8.7 Common polyphenolic compounds found in leaves of SBT (Criste et al. 2020).

iii. Organic acid

Vitamin C (L-ascorbic acid and L-dehydroascorbic acid) is a micronutrient, but it is present in large quantities ranging from 1.4 to 30.0 g/kg (Ranjith et al. 2006). SBT contains predominantly malic acid (11–60 mg/L of juice), quinic acid (7–49 mg/L of juice), citric acid (up to 0.3% of all acids), tartaric acid (up to 0.013% of all acids), and succinic acid (up to 0.6% of all acids) (Figure 8.8).

iv. Major nutrients

Lipid and Fatty Acids

Majorly, it is composed of seeds (nearly 25%), pulp (nearly 70%), and skin. There is a considerable amount of oil in fruits, which is derived either from seed or pulp (García 2019). SBT pulp oil has the dominant proportion of saturated fatty acids (palmitic acid, nearly 50%) and 52% of unsaturated fatty acids, such as palmitoleic major contributors followed by oleic (18%), linoleic (4%), and linolenic acids (2%) (Li & Beveridge 2003). The seed oil contains C18-type unsaturated fatty acids, such as linoleic acid (18:2) and linolenic acid (18:3). The considerably high content of palmitoleic acid is uncommon and distinguishes the oils from SBT pulp/peel from those from SBT seeds.

Carbohydrates

The dry matter of SBT fruit contains 400–600 g of carbohydrates per kg dry weight. SBT is also an intriguing nutraceutical due to its diverse sugar derivative composition. In SBT berries, sugar alcohols (polyols)

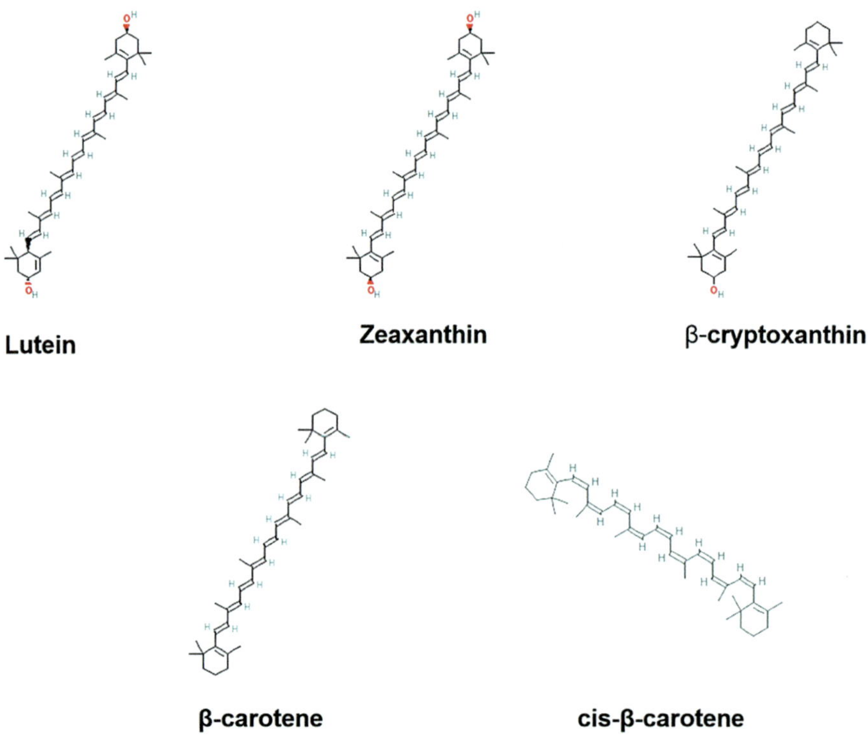

FIGURE 8.8 Common carotenoids present in berries of SBT.

TABLE 8.5
Amino Acid Composition of *Hippophae. rhamnoides* L.

Amino Acid	Content (mg/100 g)
Aspartic acid/asparagine	426.6
Proline	45.2
Threonine	36.8
Serine	28.1
Lysine	27.2
Valine	21.8
Alanine	21.2
Phenylalanine	20
Glutamine	19.4
Isoleucine	17.4
Glycine	16.7
Histidine	13.7
Tyrosine	13.4
Arginine	11.3
Cysteine	3.3
Methionine	2.3
Leucine	1.9
Tryptophan	0.5
Essential	155.9
Total	724.4

Source: Adapted from Gâtlan & Gutt (2021).

mannitol, sorbitol, and xylitol are also present; likewise, lower amounts of saccharose and ethyl glucose are found (Tiitinen et al. 2006). The low sugar and high acidity ratios are important parameters to make it acceptable in the consumer market. There is a variation in ratios of two that affects and influences the sensory properties of SBT berries.

Proteins and Amino Acids

The total protein content in the Indian variety of SBT varies from 46 to 129 g/kg dry weight (Solà Marsiñach et al. 2019). The berries of SBT are characterized by a relatively high content of protein as well as leaves. SBT juices have a variety of free amino acids that are essential for humans (Table 8.5). Also, one will observe turbidity in the juices due to cellular debris derived from cell membranes that have a considerable amount of protein.

8.6.2 Medicinal and Therapeutic Potential

i. Anti-bacterial and Anti-fungal properties

SBT is known to possess broad-spectrum anti-bacterial properties. It is effective against both gram-positive and gram-negative strains of bacteria.

Additionally, it also works against fungi from the *Saccharomycetaceae* family, such as *Candida albicans* or *Candida glabrata* (Sadowska et al. 2017). The inhibition activity assays are expressed as the minimum inhibitory concentration; the area of the inhibition zone depends on the nature of the solvent and strain of SB used. A broad spectrum of bacterial species exhibit inhibition expression upon exposure to variable concentrations of SBT. For example, *Bacillus cereus* responds to leaf extracts of SBT, while *Bacillus subtilis* responds to seed extracts of SBT (Yue et al. 2017). Certain disease-causing bacterial species like *Salmonella typhi*, *Shigella dysenteriae*, and *Helicobacter pylori* exhibited positive inhibition expression against SBT (Li et al. 2005; Kumar et al. 2013). The anti-viral activity against the influenza virus has been reported by SBT leaf extract (Enkhtaivan et al. 2017).

ii. Anti-inflammatory properties

There are various anti-inflammatory compounds found in SBT, like ursolic acid, oleanolic acid, and casuarinin, which are potent molecules that have anti-inflammatory and antioxidant properties (Kwon et al. 2012; Rédei et al. 2018). It has been reported that lipopolysaccharide-induced inflammation damage is regulated by the NF-kB pathway (Figure 8.9) using

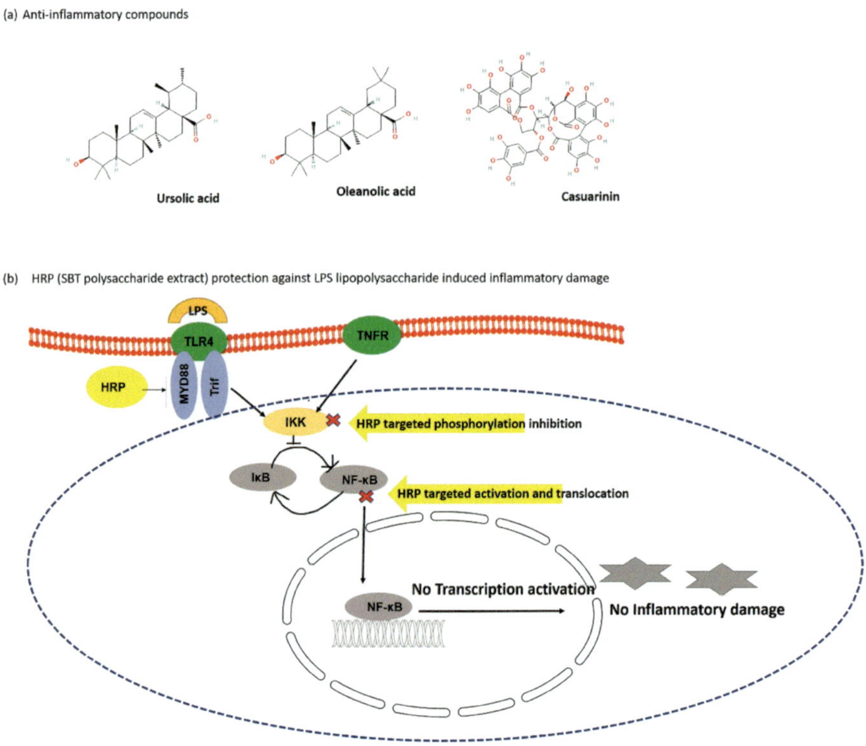

FIGURE 8.9 (a) Common anti-inflammatory compounds found in SBT (b) Polysaccharide extract of SBT (HRP) inhibit activation of inflammatory response by targeting IKK iKappa b kinase and activation and translocation of NF- kB into the nucleus (Liu et al., 2015, 2020).

polysaccharide extraction of SBT (Zhao et al. 2020; Liu et al. 2015, 2020) (Figure 8.9).

iii. Antioxidant properties

SBT is proven by both *in vitro* and *in vivo* approaches to be a very good source of antioxidants. The 2,2-diphenyl-1-picrylhydrazyl -radical-scavenging method and -Ferric Reducing Ability Of Plasma assays have suggested that seed extracts are the richest source of antioxidants (Michel et al. 2012). The phenolic compounds play an important role in the antioxidant nature of SBT (Michael et al. 2012).

8.6.3 Anti-cancerous Role (Biological and Pharmacological Activities)—In Vivo/In Vitro Studies

Various bioactive substances share structural and functional similarities with FDA-approved drugs and are a potential target of biomedical research. The phytopharmaceutical compounds, viz., phenolic compounds, are known to possess anti-cancer properties. Numerous studies demonstrate that SBT has anti-cancer properties. Various carcinogenic cell line models have shown that SBT extracts have anti-cancer effects. The anti-proliferative activity was reported from berry extracts of SBT on adenocarcinoma of stomach (AGS), colorectal (PC3), prostate (Caco-2), and mammary gland (MDA-MB 231, ACF-7) (Boivin et al. 2007). In the breast carcinoma cell lines Bcap-37 and MDA-MB 231, there was a change in the expression of apoptosis genes due to flavonoid and procyanidin extracted from the SBT seeds (Zhang et al. 2005; Wang et al. 2014). Additionally, procyanidin also inhibits fatty acid synthase biosynthesis in the MDA-MB231 breast carcinoma line (Wang et al. 2014). A leaf extract of SBT rich in catechin and phenolic compounds prevents proliferation in the rat C6 glioma cell line by inducing early apoptosis (Figure 8.10). It has been reported that human acute myeloid leukaemia cell lines (KG 1a, HL 60, and U937) have a dose-dependent effect of SBT leaf extracts on the cell cycle deceleration and induction of apoptosis. Additionally, it also inhibits microsomal lipid peroxidation (Zhamanbaeva et al. 2014, 2016).

To validate the *in vitro* studies, the *in vivo* rodent model has been explored to understand the effect of SBT extract. The SBT is rich in phenolics, flavonoids, and antioxidant compounds that are attributed to its pharmacological properties. In mice model, it has been reported that SBT increases the DNA-binding activity of - Interferon Regulatory Factor 1, interferon regulatory factor 1, an anti-oncogenic transcription factor, and causes growth suppression and apoptosis in skin tumorigenesis (Padmavathi et al. 2005).

8.6.4 Radioprotective Ability

It has been reported by Goel and Agarwal that *H. rhamnoides* berry whole extracts, denoted as RH-3, at a concentration of 25–35 mg/kg body wt., have a potential role in protecting against radiation damage, as shown by a bone marrow radiation

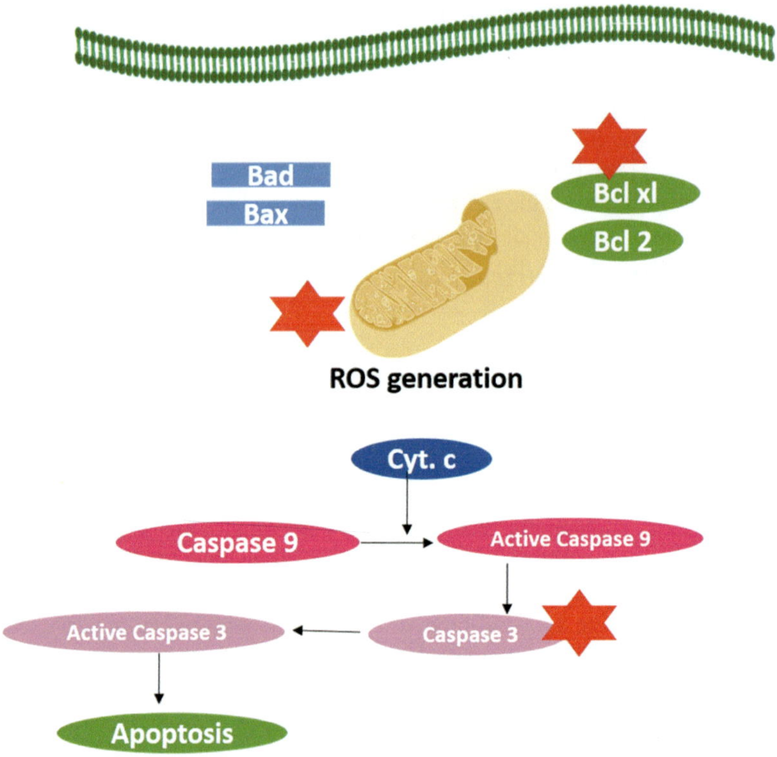

FIGURE 8.10 Apoptosis pathway and mitochondrial targets of SBT-extracted compounds.

experiment in a rodent model (Agrawala & Goel 2002). Biochemically, RH-3 interferes with Fenton's reaction, which otherwise results in the generation of powerful oxidizing agents (Goel et al. 2002). The radioprotective dose of RH-3 inhibits DNA damage and compaction of chromatin at given concentrations, as mentioned in reports, and affects caspase-3 activity, thus targeting apoptosis as well (Prem Kumar et al. 2002; Goel et al. 2003, 2004).

Additionally, it is demonstrated in the mice model that the leaf extract of *H. rhamnoides* (labelled as SBL-1) has a radioprotective role against cobalt-60 γ (^{60}Co-γ)-irradiation (Bala et al. 2009). SBL-1 has quercetin dihydrate, rutin trihydrate, and gallic acid ethyl ester as major components (Bala and Saini 2013). It is known to counter oxidative stress and affect Reactive Oxygen Species generation along with the apoptosis pathway (Bala et al. 2015, 2017). Radiation-induced "counter taste aversion"-like assays are being conducted to address the behavioural effects of these compounds for the evaluation of drugs to be used as counter medicines. SOD1 binding ligand 1 pre-treatment exhibits neuroprotection against damages caused by brain injuries (Bala et al. 2017).

8.6.5 Conclusion

SBT is characterized by a unique composition of bioactive compounds. The plant has medicinal and nutraceutical importance. It is a rich source of antioxidants, flavonoids, tannins, vitamins, carotenoids, and phenolic compounds. These compounds make it a special candidate to act as an anti-inflammatory, anti-cancer, and antioxidant plant. However, the mechanism underlying the role of SBT is yet to be explored in detail.

This chapter highlights the potential therapeutic role of quinoa and SBT species in India. Underutilized crops thus play an important role in their therapeutic properties due to their significant nutritional and medicinal value and can be viewed as future horticultural resources that will help nations ensure nutrition and food security in addition to their recreational, social, and environmental resources.

REFERENCES

Abderrahim, Fatima, Elizabeth Huanatico, Roger Segura, Silvia Arribas, M. Carmen Gonzalez, and Luis Condezo-Hoyos. "Physical features, phenolic compounds, betalains and total antioxidant capacity of coloured quinoa seeds (Chenopodium quinoa Willd.) from Peruvian Altiplano." *Food Chemistry* 183 (2015): 83–90.

Adamczewska-Sowińska, Katarzyna, Józef Sowiński, and Anna Jama-Rodzeńska. "The effect of sowing date and harvest time on leafy greens of quinoa (Chenopodium quinoa willd.) yield and selected nutritional parameters." *Agriculture* 11, no. 5 (2021): 405.

Agrawala, Paban Kumar, and Harish ChandraGoel. "Protective effect of RH-3 with special reference to radiation induced micronuclei in mouse bone marrow." (2002).

Aguilar-Tuesta, Silvana, Wile Mamani-Navarro, Clara Espinoza-Silva, Jaime Basilio-Atencio, and Luis Condezo-Hoyos. "Microencapsulated betacyanin from colored organic quinoa (Chenopodium quinoa Willd.): optimization, physicochemical characterization and accelerated storage stability." *Journal of the Science of Food and Agriculture* 98, no. 15 (2018): 5873–5883.

Alvarez-Jubete, Laura, Arendt, Elke K., and Gallagher, Eileen. "Nutritive value of pseudocereals and their increasing use as functional gluten-free ingredients." *Trends in Food Science & Technology* 21, no. (2) (2010): 106–113.

Anthony, Tracy G., Emily T. Mirek, Albert Raouf Bargoud, Lindsey Phillipson-Weiner, Christopher M. DeOliveira, Berish Wetstein, Brittany L. Graf, Peter E. Kuhn, and Ilya Raskin. "Evaluating the effect of 20-hydroxyecdysone (20HE) on mechanistic target of rapamycin complex 1 (mTORC1) signaling in the skeletal muscle and liver of rats." *Applied Physiology, Nutrition, and Metabolism* 40, no. 12 (2015): 1324–1328.

Arimboor, Ranjith and Arumughan, Chirayath. "HPLC-DAD-MS/MS profiling of antioxidant flavonoid glycosides in sea buckthorn (Hippophae rhamnoides L.) seeds." *International Journal of Food Sciences and Nutrition* 63, no. 6 (2012): 730–738.

Atefi, Masoumeh, Solmaz Mirzamohammadi, Mina Darand, and Mohammad Javad Tarrahi. "Meta-analysis of the effects of quinoa (Chenopodium quinoa) interventions on blood lipids." *Journal of Herbal Medicine* 34 (2022): 100571.

Bala, Madhu, Manish Gupta, Manu Saini, M. Z. Abdin, and Jagdish Prasad. "Sea buckthorn leaf extract protects jejunum and bone marrow of 60cobalt-gamma-irradiated mice by regulating apoptosis and tissue regeneration." *Evidence-Based Complementary and Alternative Medicine* (2015): 765705.

Bala, Madhu, Vanita Gupta, and Jagdish Prasad. "A standardized Hippophae extract (SBL-1) counters neuronal tissue injuries and changes in neurotransmitters: implications in radiation protection." *Pharmaceutical Biology* 55, no. 1 (2017): 1833–1842.

Bala, Madhu, Prasad, Jagdish, Singh, Sunil, Tiwari, Shashi, and Sawhney, Rajendra Chandra. "Whole-body radioprotective effects of SBL-1: a preparation from leaves of Hippophae rhamnoides." *Journal of Herbs, Spices & Medicinal Plants* 15, no. 2 (2009): 203–215.

Bala, Madhu, and Manu Saini. "Validated HPTLC methods for quantification of marker compounds in aqueous extract of Hippophae rhamnoides leaves." *International Journal of Pharmaceutical Sciences Review and Research* 23, no. 2 (2013): 58–63.

Bastidas, Enrique Guillermo, Roura, Raúl, Rizzolo, Daniel Alejandro Diego, Massanés, Toni, and Gomis, Ramon. "Quinoa (Chenopodium quinoa Willd), from nutritional value to potential health benefits: an integrative review." *Journal of Nutrition & Food Sciences* 6, no. 3 (2016).

Báthori, Mária, Noémi Tóth, Attila Hunyadi, Árpád Márki, and Erno Zador. "Phytoecdysteroids and anabolic-androgenic steroids-structure and effects on humans." *Current Medicinal Chemistry* 15, no. 1 (2008): 75–91.

Bazile, Didier. "Le quinoa, les enjeux d'une conquête." *Le Quinoa, Les Enjeux D'une Conquête* (2015): 1–112.

Bhargava, Anil, Sudhir, Sandeep, and Ohri, Deepak. "Quinoa (Chenopodium quinoa Willd.). An Indian perspective." *Industrial Crops and Products* 23 (2006): 73–87.

Boivin, Dominique, Melanie Blanchette, Stephane Barrette, Albert Moghrabi, and Richard Beliveau. "Inhibition of cancer cell proliferation and suppression of TNF-induced activation of NFκB by edible berry juice." *Anticancer Research* 27, no. 2 (2007): 937–948.

Borjac, Jamilah. "Antiproliferative effects of some medicinal plants on colorectal cancer: a mini review." *BAU Journal-Health and Wellbeing* 5, no. 1 (2022): 5.

Camaggio, Gigliola, and Vera Amicarelli. "The ancient crop of quinoa for world food security." Micniakiewicz M. & Popek S. *Commodity Science in Research and Practice-Future Trends and Challenges in Food Sector 1* (2014): 31–45.

Cao, Yanan, Liang Zou, Wei Li, Yu Song, Gang Zhao, and Yichen Hu. "Dietary quinoa (Chenopodium quinoa Willd.) polysaccharides ameliorate high-fat diet-induced hyperlipidemia and modulate gut microbiota." *International Journal of Biological Macromolecules* 163 (2020): 55–65.

Ciesarová, Zuzana, Michael Murkovic, Karel Cejpek, František Kreps, Blanka Tobolková, Richard Koplík, Elena Belajová et al. "Why is sea buckthorn (Hippophae rhamnoides L.) so exceptional? A review." *Food Research International* 133 (2020): 109170.

Cooper, Raymond. "Re-discovering ancient wheat varieties as functional foods." *Journal of Traditional and Complementary Medicine* 5, no. 3 (2015): 138–143.

Criste, Adriana, Adriana Cristina Urcan, Andrea Bunea, Flavia Roxana Pripon Furtuna, Neli Kinga Olah, Robert H. Madden, and Nicolae Corcionivoschi. "Phytochemical composition and biological activity of berries and leaves from four Romanian sea buckthorn (Hippophae rhamnoides L.) varieties." *Molecules* 25, no. 5 (2020): 1170.

Das, Saubhik. "Amaranths: the crop of great prospect." In *Amaranthus: A promising crop of future*, pp. 13–48. Springer, Singapore, 2016.

Da-Silva, Wagner S., John W. Harney, Brian W. Kim, Jing Li, Suzy DC Bianco, Alessandra Crescenzi, Marcelo A. Christoffolete, Stephen A. Huang, and Antonio C. Bianco. "The small polyphenolic molecule kaempferol increases cellular energy expenditure and thyroid hormone activation." *Diabetes* 56, no. 3 (2007): 767–776.

Dini, Irene, Gian Carlo Tenore, and Antonio Dini. "Antioxidant compound contents and antioxidant activity before and after cooking in sweet and bitter Chenopodium quinoa seeds." *LWT-Food Science and Technology* 43, no. 3 (2010): 447–451.

El Hazzam, Khadija, Jawhar Hafsa, Mansour Sobeh, Manal Mhada, Moha Taourirte, Kamal El Kacimi, and Abdelaziz Yasri. "An insight into saponins from Quinoa (Chenopodium quinoa Willd): A review." *Molecules* 25, no. 5 (2020): 1059.

Enkhtaivan, Gansukh, KM Maria John, Muthuraman Pandurangan, Ji Hoon Hur, Alain Simplice Leutou, and Doo Hwan Kim. "Extreme effects of Seabuckthorn extracts on influenza viruses and human cancer cells and correlation between flavonol glycosides and biological activities of extracts." *Saudi Journal of Biological Sciences* 24, no. 7 (2017): 1646–1656.

FAOSTAT. Quinoa area and production in the World, 2013. https://www.fao.org

Foucault, Anne-Sophie, Patrick Even, René Lafont, Waly Dioh, Stanislas Veillet, Daniel Tomé, Jean-François Huneau, Dominique Hermier, and Annie Quignard-Boulangé. "Quinoa extract enriched in 20-hydroxyecdysone affects energy homeostasis and intestinal fat absorption in mice fed a high-fat diet." *Physiology & Behavior* 128 (2014): 226–231.

Gandía-Herrero, Fernando, Josefa Escribano, and Francisco García-Carmona. "Biological activities of plant pigments betalains." *Critical Reviews in Food Science and Nutrition* 56, no. 6 (2016): 937–945.

García, Vanesa León. "The omega 7 as a health strategy for the skin and mucous membranes." *EC Nutrition* 14 (2019): 484–489.

Gâtlan, Anca-Mihaela, and Gheorghe Gutt. "Sea buckthorn in plant based diets. An analytical approach of sea buckthorn fruits composition: Nutritional value, applications, and health benefits." *International Journal of Environmental Research and Public Health* 18, no. 17 (2021): 8986.

Gawlik-Dziki, Urszula, Michał Świeca, Maciej Sułkowski, Dariusz Dziki, Barbara Baraniak, and Jarosław Czyż. "Antioxidant and anticancer activities of Chenopodium quinoa leaves extracts–in vitro study." *Food and Chemical Toxicology* 57 (2013): 154–160.

Gengatharan, Ashwini, Gary A. Dykes, and Wee Sim Choo. "Betalains: Natural plant pigments with potential application in functional foods." *LWT-Food Science and Technology* 64, no. 2 (2015): 645–649.

Goel, Harish Chandra, Prasad, Jagdish, Singh, Sunil, Sagar, Raj Kumar, Prem Kumar, Inderjit, and Sinha, Arun Kumar "Radioprotection by a herbal preparation of Hippophae rhamnoides, RH-3, against whole body lethal irradiation in mice." *Phytomedicine* 9, no. 1 (2002): 15–25.

Goel, Harish Chandra, Prem Kumar, Inderjit, Samanta, Namita, and Rana, Sandeep Virendra Singh "Induction of apoptosis in thymocytes by Hippophae rhamnoides: implications in radioprotection." *Journal of Environmental Pathology, Toxicology and Oncology* 23, no. 2 (2004).

Goel, H. C., I. Prem Kumar, Namita Samanta, and S. V. S. Rana. "Induction of DNA-protein cross-links by Hippophae rhamnoides: implications in radioprotection and cytotoxicity." *Molecular and Cellular Biochemistry* 245 (2003): 57–67.

Gómez-Caravaca, Ana Maria, Giovanna Iafelice, Vito Verardo, Emanuele Marconi, and Maria Fiorenza Caboni. "Influence of pearling process on phenolic and saponin content in quinoa (Chenopodium quinoa Willd)." *Food Chemistry* 157 (2014): 174–178.

Gómez-Caravaca, A.M., Segura-Carretero, A., Fern andezGutierrez, A. & Caboni, M.F. (2011). Simultaneous determination of phenolic compounds and saponins in quinoa (Chenopodium quinoa Willd) by a liquid chromatography–diode array detection–electrospray ionization–time-of-flight mass spectrometry methodology. *Journal of Agricultural and Food Chemistry*, 59, 10815–10825

González, Juan A., Yotaro Konishi, Marcela Bruno, Mariana Valoy, and Fernando E. Prado. "Interrelationships among seed yield, total protein and amino acid composition of ten quinoa (Chenopodium quinoa) cultivars from two different agroecological regions." *Journal of the Science of Food and Agriculture* 92, no. 6 (2012): 1222–1229.

Gordon, Edward. "Analysing the versatility and complexity of cereal grains." *Food Science Newsletters* 88 (2006): 1–5.

Gorinstein, Shela, Antonin Lojek, Milan Číž, Elke Pawelzik, Efren Delgado-Licon, Oscar J. Medina, Marlén Moreno, Ines Arnao Salas, and Ivan Goshev. "Comparison of composition and antioxidant capacity of some cereals and pseudocereals." *International Journal of Food Science & Technology* 43, no. 4 (2008): 629–637.

Graf, Brittany L., Alexander Poulev, Peter Kuhn, Mary H. Grace, Mary Ann Lila, and Ilya Raskin. "Quinoa seeds leach phytoecdysteroids and other compounds with anti-diabetic properties." *Food Chemistry* 163 (2014): 178–185.

Graf, Brittany L., Leonel E. Rojo, Jose Delatorre-Herrera, Alexander Poulev, Camila Calfio, and Ilya Raskin. "Phytoecdysteroids and flavonoid glycosides among Chilean and commercial sources of Chenopodium quinoa: Variation and correlation to physico-chemical characteristics." *Journal of the Science of Food and Agriculture* 96, no. 2 (2016): 633–643.

Graf, Brittany L., Patricio Rojas-Silva, Leonel E. Rojo, Jose Delatorre-Herrera, Manuel E. Baldeón, and Ilya Raskin. "Innovations in health value and functional food development of quinoa (Chenopodium quinoa Willd.)." *Comprehensive Reviews in Food Science and Food Safety* 14, no. 4 (2015): 431–445.Guo, Ruixue, Xinbo Guo, Tong Li, Xiong Fu, and Rui Hai Liu. "Comparative assessment of phytochemical profiles, antioxidant and antiproliferative activities of Sea buckthorn (Hippophaë rhamnoides L.) berries." *Food Chemistry* 221 (2017): 997–1003.

Han, Yameng, Jianwei Chi, Mingwei Zhang, Ruifen Zhang, Sanhong Fan, Lihong Dong, Fei Huang, and Lei Liu. "Changes in saponins, phenolics and antioxidant activity of quinoa (Chenopodium quinoa willd) during milling process." *LWT* 114 (2019): 108381.

Harborne, Jeffrey B., and Christine A. Williams. "Advances in flavonoid research since 1992." *Phytochemistry* 55, no. 6 (2000): 481–504.

Hemalatha, P., Dikki Pedenla Bomzan, BV Sathyendra Rao, and Yadahally N. Sreerama. "Distribution of phenolic antioxidants in whole and milled fractions of quinoa and their inhibitory effects on α-amylase and α-glucosidase activities." *Food Chemistry* 199 (2016): 330–338.

HernándezLedesma, Blanca. "Quinoa (Chenopodium quinoa Willd.) as source of bioactive compounds: A review." *Bioactive Compounds in Health and Disease* 2, no. 3 (2019): 27–47.

Hosny, Heba, Nayra Omran, and Heba Handoussa. "Edible seeds with potential anti-obesity impact: A Review." *International Journal of Plant Based Pharmaceuticals* 2, no. 1 (2022): 64–81.

Hu, Yichen, Jinming Zhang, Liang Zou, Chaomei Fu, Peng Li, and Gang Zhao. "Chemical characterization, antioxidant, immune-regulating and anticancer activities of a novel bioactive polysaccharide from Chenopodium quinoa seeds." *International Journal of Biological Macromolecules* 99 (2017): 622–629.

Hussain, M. Iftikhar, Muhammad Farooq, Qamar Abbas Syed, Anum Ishaq, Abdullah Ahmed Al-Ghamdi, and Ashraf A. Hatamleh. "Botany, nutritional value, phytochemical composition and biological activities of quinoa." *Plants* 10, no. 11 (2021): 2258.

Ibrahem, Eman S., and Ghada H. Mohamed. "Effect of different preparatory processes on improving the nutritional value of quinoa (Chenopodium quinoa, Willd) and studying its effect on obese anemic rats." *Egyptian Journal of Nutrition and Health* 16, no. 2 (2021): 71–94.

Jacobsen, Soren Einar. "The situation for quinoa and its production in southern Bolivia: from economic success to environmental disaster." *Journal of Agronomy and Crop Science* 197, no. 5 (2011): 390–399.

Kadereit, Gerd, Borsch, Thomas, Weising, Kurt, and Freitag, Hans. "Phylogeny of Amaranthaceae and Chenopodiaceae and the evolution of C4 photosynthesis." *International Journal of Plant Sciences* 164, no. 6 (2003): 959–986.

Kumar, M. S. Yogendra, R. J. Tirpude, D. T. Maheshwari, Anju Bansal, and Ksipra Misra. "Antioxidant and antimicrobial properties of phenolic rich fraction of Seabuckthorn (Hippophae rhamnoides L.) leaves in vitro." *Food Chemistry* 141, no. 4 (2013): 3443–3450.

Kumpun, Saowanee, Annick Maria, Sophie Crouzet, Nathalie Evrard-Todeschi, Jean-Pierre Girault, and René Lafont. "Ecdysteroids from Chenopodium quinoa Willd., an ancient Andean crop of high nutritional value." *Food Chemistry* 125, no. 4 (2011): 1226–1234.

Kwon, Dong-Joo, Young-Soo Bae, Sung Mi Ju, Ah Ra Goh, Gi Soo Youn, Soo Young Choi, and Jinseu Park. "Casuarinin suppresses TARC/CCL17 and MDC/CCL22 production via blockade of NF-κB and STAT1 activation in HaCaT cells." *Biochemical and Biophysical Research Communications* 417, no. 4 (2012): 1254–1259.

Lápez-Cervantes, Jaime, Dalia Sánchez-Machado, David S. de la Mora-López, and Ana Sanches-Silva. "Quinoa (Chenopodium quinoa Willd.): exploring a superfood from Andean indigenous cultures with potential to reduce cardiovascular disease (CVD) risk markers." *Current Molecular Pharmacology* 14, no. 6 (2021): 925–934.

Li, Thomas Shih-Chung, and Beveridge, Thomas Harry James. "Sea buckthorn (Hippophae rhamnoides L.): production and utilization." In *Sea Buckthorn (Hippophae rhamnoides L.): Production and Utilization*, edited by Thomas Shih-Chung Li and Thomas Harry James Beveridge, pp. 101–106. NRC Research Press, 2003.

Li, Yang, Chen Xu, Qiang Zhang, Jun Yan Liu, and Ren Xiang Tan. "In vitro anti-Helicobacter pylori action of 30 Chinese herbal medicines used to treat ulcer diseases." *Journal of Ethnopharmacology* 98, no. 3 (2005): 329–333.

Lim, Jeong Gyu, Hyun-Mee Park, and Ki Sun Yoon. "Analysis of saponin composition and comparison of the antioxidant activity of various parts of the quinoa plant (Chenopodium quinoa Willd.)." *Food Science & Nutrition* 8, no. 1 (2020): 694–702.

Lin, Minyi, Peipei Han, Yuying Li, Weixuan Wang, Daowan Lai, and Ligang Zhou. "Quinoa secondary metabolites and their biological activities or functions." *Molecules* 24, no. 13 (2019): 2512.

Liu, Fengting, Tao Wang, Xiaoxia Li, Jinxue Jia, Qin Lin, and Yongzhi Xue. "Involvement of NF-κB in the reversal of CYP3A down-regulation induced by sea buckthorn in BCG-induced rats." *Plos One* 15, no. 9 (2020): e0238810.

Liu, Huan, Wei Zhang, Shichao Dong, Liang Song, Shimin Zhao, Chunyan Wu, Xue Wang et al. "Protective effects of sea buckthorn polysaccharide extracts against LPS/d-GalN-induced acute liver failure in mice via suppressing TLR4-NF-κB signaling." *Journal of Ethnopharmacology* 176 (2015): 69–78.

Lutz, Mariane, Angelica Martínez, and Enrique A. Martínez. "Daidzein and Genistein contents in seeds of quinoa (Chenopodium quinoa Willd.) from local ecotypes grown in arid Chile." *Industrial Crops and Products* 49 (2013): 117–121.

Ma, Xueying, Wei Yang, Heikki Kallio, and Baoru Yang. "Health promoting properties and sensory characteristics of phytochemicals in berries and leaves of sea buckthorn (Hippophaë rhamnoides)." *Critical Reviews in Food Science and Nutrition* 62, no. 14 (2022): 3798–3816.

Madadi, Elaheh, Sahand Mazloum-Ravasan, Jae Sik Yu, Ji Won Ha, Hamed Hamishehkar, and Ki Hyun Kim. "Therapeutic application of betalains: A review." *Plants* 9, no. 9 (2020): 1219.

McKeown, Nicola M., Paul F. Jacques, Chris J. Seal, Jan de Vries, Satya S. Jonnalagadda, Roger Clemens, Densie Webb et al. "Whole grains and health: from theory to practice—highlights of the Grains for Health Foundation's Whole Grains Summit 2012." *The Journal of Nutrition* 143, no. 5 (2013): 744S–758S.

Melini, Valentina, and Francesca Melini. "Functional components and anti-nutritional factors in gluten-free grains: A focus on quinoa seeds." *Foods* 10, no. 2 (2021): 351.

Michel, Thomas, Emilie Destandau, Gaëtan Le Floch, Marie Elisabeth Lucchesi, and Claire Elfakir. "Antimicrobial, antioxidant and phytochemical investigations of sea buckthorn (Hippophaë rhamnoides L.) leaf, stem, root and seed." *Food Chemistry* 131, no. 3 (2012): 754–760.

Miranda, Margarita, Antonio Vega-Gálvez, Enrique Martinez, Jéssica López, María José Rodríguez, Karem Henríquez, and Francisco Fuentes. "Genetic diversity and comparison of physicochemical and nutritional characteristics of six quinoa (Chenopodium quinoa Willd.) genotypes cultivated in Chile." *Food Science and Technology* 32 (2012): 835–843.

Mohamed Ahmed, Isam A., Fahad Al Juhaimi, and Mehmet Musa Özcan. "Insights into the nutritional value and bioactive properties of quinoa (Chenopodium quinoa): past, present and future prospective." *International Journal of Food Science & Technology* 56, no. 8 (2021): 3726–3741.

Mohamed, Doha A., Karem Aly Fouda, and Rasha S. Mohamed. "In vitro anticancer activity of quinoa and safflower seeds and their preventive effects on non-alcoholic fatty liver." *Pakistan Journal of Biological Sciences: PJBS* 22, no. 8 (2019): 383–392.

Moreau, Robert A., Laura Nyström, Bruce D. Whitaker, Jill K. Winkler-Moser, David J. Baer, Sarah K. Gebauer, and Kevin B. Hicks. "Phytosterols and their derivatives: Structural diversity, distribution, metabolism, analysis, and health-promoting uses." *Progress in Lipid Research* 70 (2018): 35–61.

Navarro-Perez, Diana, Jessica Radcliffe, Audrey Tierney, and Markandeya Jois. "Quinoa seed lowers serum triglycerides in overweight and obese subjects: a dose-response randomized controlled clinical trial." *Current Developments in Nutrition* 1, no. 9 (2017): e001321.

Navruz-Varli, Semra, and Nevin Sanlier. "Nutritional and health benefits of quinoa (Chenopodium quinoa Willd.)." *Journal of Cereal Science* 69 (2016): 371–376.

Nsimba, Robert Yawadio, Hiroe Kikuzaki, and Yotaro Konishi. "Ecdysteroids act as inhibitors of calf skin collagenase and oxidative stress." *Journal of Biochemical and Molecular Toxicology* 22, no. 4 (2008): 240–250.

Ogungbenle, H. N. "Nutritional evaluation and functional properties of quinoa (Chenopodium quinoa) flour." *International Journal of Food Sciences and Nutrition* 54, no. 2 (2003): 153–158.

Olas, Beata. "Sea buckthorn as a source of important bioactive compounds in cardiovascular diseases." *Food and Chemical Toxicology* 97 (2016): 199–204.

Olas, Beata, Bogdan Kontek, Paulina Malinowska, Jerzy Żuchowski, and Anna Stochmal. "Hippophae rhamnoides L. fruits reduce the oxidative stress in human blood platelets and plasma." *Oxidative Medicine and Cellular Longevity* 2016 (2016).

Olas, Beata, Bartosz Skalski, and Karolina Ulanowska. "The anticancer activity of sea buckthorn [Elaeagnus rhamnoides (L.) A. Nelson]." *Frontiers in Pharmacology* 9 (2018): 232.

Olsson, Marie E., Karl-Erik Gustavsson, Staffan Andersson, Åke Nilsson, and Rui-Dong Duan. "Inhibition of cancer cell proliferation in vitro by fruit and berry extracts and correlations with antioxidant levels." *Journal of Agricultural and Food Chemistry* 52, no. 24 (2004): 7264–7271.

Padmavathi, Bandhuvula, Meenakshi Upreti, Virendra Singh, A. Ramesha Rao, Rana P. Singh, and Pramod C. Rath. "Chemoprevention by Hippophae rhamnoides: effects on tumorigenesis, phase II and antioxidant enzymes, and IRF-1 transcription factor." *Nutrition and Cancer* 51, no. 1 (2005): 59–67.

Park, Jin Hwa, Yun Jin Lee, Yeon Ho Kim, and Ki Sun Yoon. "Antioxidant and antimicrobial activities of Quinoa (Chenopodium quinoa Willd.) seeds cultivated in Korea." *Preventive Nutrition and Food Science* 22, no. 3 (2017): 195.

Paśko, Paweł, Henryk Bartoń, Paweł Zagrodzki, Shela Gorinstein, Maria Fołta, and Zofia Zachwieja. "Anthocyanins, total polyphenols and antioxidant activity in amaranth and quinoa seeds and sprouts during their growth." *Food Chemistry* 115, no. 3 (2009): 994–998.

Paśko, Paweł Andrzej, Sajewicz, Małgorzata, Gorinstein, Svetlana, and Zachwieja, Zbigniew. "Analysis of selected phenolic acids and flavonoids in Amaranthus cruentus and Chenopodium quinoa seeds and sprouts by HPLC." *Acta Chromatographica* 20, no. 4 (2008): 661–672.

Pathan, Safiullah, Frieda Eivazi, Babu Valliyodan, Kamalendu Paul, Grato Ndunguru, and Kerry Clark. "Nutritional composition of the green leaves of quinoa (Chenopodium quinoa Willd.)." *Journal of Food Research* 8, no. 6 (2019): 55–65.

Pathan, Safiullah, and Rafat A. Siddiqui. "Nutritional composition and bioactive components in quinoa (Chenopodium quinoa Willd.) greens: A review." *Nutrients* 14, no. 3 (2022): 558.

Pellegrini, Marika, Raquel Lucas-Gonzalez, Juana Fernández-López, Antonella Ricci, José A. Pérez-Álvarez, Claudio Lo Sterzo, and Manuel Viuda-Martos. "Bioaccessibility of polyphenolic compounds of six quinoa seeds during in vitro gastrointestinal digestion." *Journal of Functional Foods* 38 (2017): 77–88.

Pourshahidi, L. Kirsty, Eduardo Caballero, Alejandro Osses, Barry W. Hyland, Nigel G. Ternan, and Chris IR Gill. "Modest improvement in CVD risk markers in older adults following quinoa (*Chenopodium quinoa* Willd.) consumption: a randomized-controlled crossover study with a novel food product." *European Journal of Nutrition* 59, no. 7 (2020): 3313–3323.

Prem Kumar, Inderjit, Samanta, Namita, and Goel, Harish Chandra. "Modulation of chromatin organization by RH-3, a preparation of Hippophae rhamnoides, a possible role in radioprotection." *Molecular and Cellular Biochemistry* 238 (2002): 1–9.

Ramesh, Kottala, Suneetha Devi, Kanithi Bala, Gopinath, Kandasamy Arunachalam, and Uma Devi, Manubolu "Growth, yield and economics of quinoa as influenced by different dates of sowing and varied crop geometry." *International Journal of Pure & Applied Bioscience* 5, no. 6 (2017): 849–854.

Ranjith, Arumughan, Sarin Kumar, Karingannoor, Venugopalan, Vazhoor Venugopalan, Arumughan, Chirayath, Sawhney, Rajendra Chandra, and Virendra Singh. "Fatty acids, tocols, and carotenoids in pulp oil of three sea buckthorn species (Hippophae rhamnoides, H. salicifolia, and H. tibetana) grown in the Indian Himalayas." *Journal of the American Oil Chemists' Society* 83, no. 4 (2006): 359–364.

Rédei, Dóra, Norbert Kúsz, Nikoletta Jedlinszki, Gábor Blazsó, István Zupkó, and Judit Hohmann. "Bioactivity-guided investigation of the anti-inflammatory activity of Hippophae rhamnoides fruits." *Planta Medica* 84, no. 1 (2018): 26–33.

Renard, Catherine M. G. C., Gundolf Wende, and Elaine J. Booth. "Cell wall phenolics and polysaccharides in different tissues of quinoa (Chenopodium quinoa Willd)." *Journal of the Science of Food and Agriculture* 79, no. 14 (1999): 2029–2034.

Repo-Carrasco-Valencia, Ritva Ann-Mari, and Lesli Astuhuaman Serna. "Quinoa (Chenopodium quinoa, Willd.) as a source of dietary fiber and other functional components." *Food Science and Technology* 31 (2011): 225–230.

Repo-Carrasco-Valencia, Ritva, Jarkko K. Hellström, Juha-Matti Pihlava, and Pirjo H. Mattila. "Flavonoids and other phenolic compounds in Andean indigenous grains: Quinoa (Chenopodium quinoa), kañiwa (Chenopodium pallidicaule) and kiwicha (Amaranthus caudatus)." *Food Chemistry* 120, no. 1 (2010): 128–133.

Sadowska, Beata, Aleksandra Budzyńska, Anna Stochmal, Jerzy Żuchowski, and Barbara Różalska. "Novel properties of Hippophae rhamnoides L. twig and leaf extracts-anti-virulence action and synergy with antifungals studied in vitro on Candida spp. model." *Microbial Pathogenesis* 107 (2017): 372–379.

Salminen, Juha-Pekka, and Maarit Karonen. "Chemical ecology of tannins and other phenolics: we need a change in approach." *Functional Ecology* 25, no. 2 (2011): 325–338.

Samtiya, Mrinal, Rotimi E. Aluko, Tejpal Dhewa, and José Manuel Moreno-Rojas. "Potential health benefits of plant food-derived bioactive components: An overview." *Foods* 10, no. 4 (2021): 839.

Shewry, Peter R. "The major seed storage proteins of spelt wheat, sorghum, millets and pseudocereals." In *Pseudocereals and less common cereals*, pp. 1–24. Springer, 2002.

Silva, Vanessa Oliveira, Michele Lima Gregório, Daniel Fernando Villafanha, and Moacir Fernandes de Godoy. "Effects of Intake of Processed Quinoa Seeds on Lipid Profile in Patients with Coronary Heart Disease." *International Journal of Sciences* 7, no. 3 (2018): 8–14.

Singh, Kunwar Vikash, and Rashmi Singh. "Quinoa (Chenopodium quinoa Willd), functional superfood for today's world: A Review." *World Scientific News* 58 (2016): 84–96.

Solà Marsiñach, Marta, and Aleix Pellejero Cuenca. "The impact of sea buckthorn oil fatty acids on human health." *Lipids in Health and Disease* 18, no. 1 (2019): 1–11.

Srinivasa Rao, K. Sarikotha panta quinoa, *Sakhi News*, 10 on 11.08.2015.

Suomela, Jukka-Pekka, Ritva Repo-Carrasco-Valencia, and Mariane Lutz. "Andean native grains, quinoa, and lupin as sources of bioactive components." In *Native crops in Latin America*, edited by Patricia Vit, pp. 1–34. CRC Press, 2022.

Suryakumar, Geetha, and Asheesh Gupta. "Medicinal and therapeutic potential of Sea buckthorn (Hippophae rhamnoides L.)." *Journal of Ethnopharmacology* 138, no. 2 (2011): 268–278.

Tang, Yao, Xihong Li, Peter X. Chen, Bing Zhang, Marta Hernandez, Hua Zhang, Massimo F. Marcone, Ronghua Liu, and Rong Tsao. "Characterisation of fatty acid, carotenoid, tocopherol/tocotrienol compositions and antioxidant activities in seeds of three Chenopodium quinoa Willd. genotypes." *Food Chemistry* 174 (2015b): 502–508.

Tang, Yao, Xihong Li, Bing Zhang, Peter X. Chen, Ronghua Liu, and Rong Tsao. "Characterisation of phenolics, betanins and antioxidant activities in seeds of three Chenopodium quinoa Willd. genotypes." *Food Chemistry* 166 (2015a): 380–388.

The World's Healthiest Foods. Quinoa. Accessed at https://whfoods.org//genpage.php?tname=foodspice&dbid=142

Tiitinen, Katja, Marjatta Vahvaselkä, Mari Hakala, Simo Laakso, and Heikki Kallio. "Malolactic fermentation in sea buckthorn (Hippophaë rhamnoides L.) juice processing." *European Food Research and Technology* 222 (2006): 686–691.

Tsao, Rong. "Chemistry and biochemistry of dietary polyphenols." *Nutrients* 2, no. 12 (2010): 1231–1246.

U.S. Department of Agriculture, Agricultural Research Service. (2013). *USDA national nutrient database for standard reference, release 26*. Nutrient Data Laboratory, Beltsville Human Nutrition Research Center, Agriculture Research Service, United States Department of Agriculture. Retrieved from Nutrient Data Laboratory Home Page, http://www.ars.usda.gov/ba/bhnrc/ndl

Valenzuela Zamudio, Francisco, and Maira Rubi Segura Campos. "Amaranth, quinoa and chia bioactive peptides: a comprehensive review on three ancient grains and their potential role in management and prevention of Type 2 diabetes." *Critical Reviews in Food Science and Nutrition* 62, no. 10 (2022): 2707–2721.

Vega-Gálvez, Antonio, Margarita Miranda, Judith Vergara, Elsa Uribe, Luis Puente, and Enrique A. Martínez. "Nutrition facts and functional potential of quinoa (Chenopodium quinoa willd.), an ancient Andean grain: a review." *Journal of the Science of Food and Agriculture* 90, no. 15 (2010): 2541–2547.

Villacrés, Elena, María Quelal, Susana Galarza, Diana Iza, and Edmundo Silva. "Nutritional value and bioactive compounds of leaves and grains from Quinoa (Chenopodium quinoa Willd.)." *Plants* 11, no. 2 (2022): 213.

Wang, Yi, Fangyuan Nie, Jian Ouyang, Xiaoyan Wang, and Xiaofeng Ma. "Inhibitory effects of sea buckthorn procyanidins on fatty acid synthase and MDA-MB-231 cells." *Tumor Biology* 35 (2014): 9563–9569.

Woldemichael, Girma Moges, and Michael Wink. "Identification and biological activities of triterpenoid saponins from Chenopodium quinoa." *Journal of Agricultural and Food Chemistry* 49, no. 5 (2001): 2327–2332.

Xia, Xichao, Qingyuan Zhang, Rongzhi Liu, Zhongxiao Wang, Nianya Tang, Fei Liu, Guosheng Huang et al. "Effects of 20-hydroxyecdysone on improving memory deficits in streptozotocin-induced type 1 diabetes mellitus in rat." *European Journal of Pharmacology* 740 (2014): 45–52.

Yendo, Anna CA, Fernanda de Costa, Grace Gosmann, and Arthur G. Fett-Neto. "Production of plant bioactive triterpenoid saponins: elicitation strategies and target genes to improve yields." *Molecular Biotechnology* 46, no. 1 (2010): 94–104.

Yue, Xuan-Feng, Xiao Shang, Zhi-Juan Zhang, and Yan-Ni Zhang. "Phytochemical composition and antibacterial activity of the essential oils from different parts of sea buckthorn (Hippophae rhamnoides L.)." *Journal of Food and Drug Analysis* 25, no. 2 (2017): 327–332.

Zevallos, Victor F., Irene L. Herencia, Fuju Chang, Suzanne Donnelly, Julia H. Ellis, and Paul J. Ciclitira. "Gastrointestinal effects of eating quinoa (Chenopodium quinoaWilld.) in celiac patients." *Official Journal of the American College of Gastroenterology | ACG* 109, no. 2 (2014): 270–278.

-Zhamanbaeva, Gulnara Tuleukhanovna, Aralbayeva, Asel Nursultanovna, Murzakhmetova, Maira Kuderbekovna, Tuleukhanov, Sabit Tuleukhanovich, and Danilenko, Mikhail Petrovich. "Cooperative antiproliferative and differentiation-enhancing activity of medicinal plant extracts in acute myeloid leukemia cells." *Biomedicine & Pharmacotherapy* 82 (2016): 80–89.

Zhamanbaeva, Gulnara Tuleukhanovna, Murzakhmetova, Maira Kuderbekovna, Tuleukhanov, Sabit Tuleukhanovich, and Danilenko, Mikhail Petrovich. "Antitumor activity of ethanol extract from Hippophae rhamnoides L. leaves towards human acute myeloid leukemia cells in vitro." *Bulletin of Experimental Biology and Medicine* 158 (2014): 252–255.

Zhang, Ping, Yu-Chang Mao, Bin Sun, Min Qian, and Wei-Jing Qu. "Changes in apoptosis-related genes expression profile in human breast carcinoma cell line Bcap-37 induced by flavonoids from seed residues of Hippophae Rhamnoides L." *Ai Zheng = Aizheng = Chinese Journal of Cancer* 24, no. 4 (2005): 454–460.

Zhao, Lei, Muyang Li, Kecheng Sun, Shuai Su, Tingting Geng, and Hui Sun. "Hippophae rhamnoides polysaccharides protect IPEC-J2 cells from LPS-induced inflammation, apoptosis and barrier dysfunction in vitro via inhibiting TLR4/NF-κB signaling pathway." *International Journal of Biological Macromolecules* 155 (2020): 1202–1215.

9 Neglected and Underutilized Crops
Towards Drug Discovery for Cancer

Hussaina Banu Buhari Malkhan Ali
Abu Dhabi, United Arab Emirates

9.1 MILLETS

Millets are one of the indigenous and traditionally significant cereal crops, cultivated for their coarse grains of diverse nutraceutical and therapeutical values. Archaeobotanical records report the evidence of cultivation of millets across Eurasia, prior to the early Neolithic period or new stone age, and had been utilized as food, feed and forage throughout the human history [1]. They are the sixth most cultivated cereal grain crop in the agricultural production of the world. Several millets are consumed as a staple food for centuries by larger segments of Asian and African population, and they are made into traditional foods and dishes such as flat-breads, porridge and dumplings, prepared mostly from the whole meal.

Millets are good sources of protein, carbohydrates, vitamins, minerals, dietary fibre and bioactive compounds such as polyphenols and flavonoids, and the regular consumption of millets results in the management of a healthy weight, delaying ageing and prevention of onset of diseases and disorders such as diabetes, cardiovascular conditions, inflammatory diseases and cancer. This chapter reviews the immense nutritional and health benefits of millets, mainly the roles of different variety of millets, rich in various phytochemicals associated with the prevention and treatment of various cancers.

9.2 DRAUGHT-RESISTANT AND DISEASE-RESISTANT COVER CROP

Millets belong to the family of grasses, broadly grown as non-commodity cover crops providing soil-improving benefits in between the growing seasons of cash crops to prevent soil erosion, replenish the soil with nutrients and offer excellent resistance to pests and other diseases. They are also known as "dry crops" or "famine crops", as they can tolerate draught and can be well grown in even low rainfall lands and marginal environments with poor moisture retention, upon which the other major cereal crops cannot deliver a good yield [2–4]. In several parts of Central Africa and Asia having marginal rainfall and water shortages, millets are grown widely for their nutritional benefits and used by the indigenous people as a staple food crop.

DOI: 10.1201/9781003477792-9

Besides, being draught resistant, millets are also very well known for their disease or pest-resistant properties. When compared to the other food grains, millets are resistant to insects and pests during cultivation and storage during conditions like famine. Studies have shown the inherent structure of the grain with the thick seed coat and richness of the polyphenolic compounds, particularly flavonoids, tannins and phenolic acids such as p-coumaric acid. Also, the diterpenoid phytoalexin produced by the millet family that acts as a natural antibiotic against the plant pathogens plays a critical role in conferring the insect and fungal resistance [5–8].

9.3 MILLET VARIETIES

Millets are from the grass family of Poaceae and includes a multitude of varieties. In general, millets are classified as major or minor millets based on their grain size. The common major millets include *Setaria italic* (foxtail millet), *Pennisetum glaucum* (pearl millet), *Eleusine coracana* (finger millet), and *Panicum miliaceum* (proso/white millet) and the minor millets include *Paspalum scrobiculatum* (kodo millet), *Sorghum spp.* (sorghum/great millet), *Brachiaria deflexa* (Guinea millet), *Urochloa ramose* (browntop millet), *Panicum Sumatrense* (little millet) and Echinochloa spp. (barnyard millet). Due to their outstanding nutritional benefits and remarkable resilience to low rainfall and draught, they are cultivated, across the globe, for their use as food in many of the developing countries and as feed in some of the developed countries.

9.4 NEGLECTED AND UNDERUTILIZED CROP

Despite their advantages in terms of their adaptability to arid and semi-arid lands, and their potential to contribute nutrition and health, millets are the least exploited and less cultivated plant species than the other major conventional cereal crops such as rice, wheat and maize [9–11]. One of the principal causes of their limited cultivation is the lesser appreciation by the farmers, researchers and legislators due to the lack of awareness of their economic value and commercial importance, and especially their enriched nutritional benefits and therapeutic properties of the different types of millets. Many of these neglected and underutilized crop species can be greater utilized by encouraging research on nutritional composition of these crops to increase our knowledge and understanding and then integrating this knowledge into the framework of nutrition and food security policies, strategies and actions of any nation [12,13].

9.5 MILLETS: HEALTH AND NUTRITIONAL BENEFITS

In common, all of the millet varieties are abundant in various vital macroelements and microelements, such as proteins, carbohydrates, fatty acids, minerals and vitamins, dietary fibres and other bioactive compounds that confer positive effect on human health such as polyphenols. They have a very high nutritional value and offer immense potential to combat malnutrition and hunger [14–16].

Studies show that the edible part of millet comprises 60%–70% of carbohydrates; 7%–11% of proteins rich in essential amino acids, particularly leucine, lysine and methionine and 1.5%–5% of fats with mainly high content of unsaturated fatty acids such as linoleic acid [17–19]. Besides the macroelements, millets contain a plethora of micronutrients such as minerals (largely magnesium, manganese, calcium and phosphorus) and vitamins (mainly tocopherol and carotenoid) [20–23].

Besides the nutritional benefits, millet grains are also rich in several bioactive phytochemicals that can provide health benefits and improve health by reducing the risk of chronic diseases and treating life style disorders such as obesity, cardiovascular diseases, diabetes and cancer [24–26]. Furthermore, millets are commonly non-allergenic and gluten-free grains recommended for use in celiac disease conditions, wherein the ingestion of gluten results in the damage of villi of small intestine [27].

Many countries such as South Asia, Southeast Asia and Africa have the traditional practice of incorporating the food grains such as pearl millet, finger millet and sorghum in the diet of infants and growing children as complementary food due to their excellent nutrient profile. Dhliwayo et al. studied the complementary foods given to children in several African countries and observed that proso millet, Guinea millet, finger millet, sorghum and pearl millet are the most common ingredients of the traditional blend of cereal-based food grains fed to children of age between 6 and 23 months in African countries like Nigeria, Uganda, Ethiopia, Ghana and Kenya [28].

9.6 MILLETS: RICH SOURCE OF ANTIOXIDANTS

Dietary antioxidant supplements are of particular interest in fighting the damage caused by free radicals. These are substances that significantly reduce the adverse damage caused by the free radicals, thereby lowering the risks of chronic stress-related diseases. Examples of common natural forms of antioxidants are ascorbic acid (vitamin C), polyphenols, beta-carotene, lycopene, zeaxanthin, lutein and tocopherol (vitamin E), which are found in fruits, vegetables, whole grains, nuts, legumes, herbs and spices.

Research indicates that eating a diet that includes millet increases the prevention of oxidative stress damage in cells, as the millets are rich in several bioactive phytochemicals in their outer bran layers or seed coat, which has antioxidant properties and will help in the prevention, management and treatment of cancer. Especially, millets are rich in antioxidative bioactive compounds that can promote health, as well as prevent and treat diseases, including phenolic compounds, especially phenolic acids, flavonoids, proanthocyanidins, tannins and also xylo-oligosaccharides, fibres, bioactive peptides, vitamin E and carotenoids [24,29–31]. Also, several studies show that fermentation and germination of millets promote and enrich the concentration of antioxidants, particularly antioxidative bioactive peptides, in most millets [32–34].

9.7 FREE RADICALS: OXIDATIVE CELLULAR DAMAGE

On a regular basis, we are increasingly exposed to several sources of free radicals such as xenobiotics (chemical substances foreign to the body), pollutants, radiation, pharmaceutical/cosmeceutical agents and processed foods, meat and oil. Basically, the free radicals are unstable atoms of compounds that possess one or more unpaired electron, rendering it highly unstable and reactive with other compounds in order to either loose or gain an electron to attain stability. The free radicals in the body affect all of the biological macromolecules of the body such as proteins, lipids and DNA, mainly accelerating the ageing process and to cause chronic health conditions such as cancer, diabetes, heart diseases, inflammatory diseases, atherosclerosis and degenerative diseases. Especially, as the body progresses with age, the ability to scavenge these free radicals very much decreases, increasing the propensity of the body to accelerated signs of ageing and occurrence of chronic diseases and disorders.

9.8 ANTIOXIDANTS: INHIBITS CELLULAR DAMAGE

Antioxidants are the ones that prevent the cellular damage resulting from oxidative stress caused by the free radicals, which are produced indigenously during metabolic reactions and/or by exposure to chemicals and radiations. The antioxidants decrease the free radicals by either inhibiting the enzymes such as Nicotinamide adenine dinucleotide phosphate (NADPH) oxidase and xanthine oxidase associated with the generation of free radicals or by increasing the levels of enzymes such as superoxide dismutase, catalase and glutathione peroxidase, which has antioxidant properties against the free radicals generated in the body [35–37].

9.9 PHENOLICS: PHENOLS AND POLYPHENOLS

The common millet varieties, mainly kodo millet, pearl millet, little millet, finger millet, foxtail millet and proso millet, are rich in several bound and unbound phenolic acids, such as ferulic acid and p-coumaric acid, and possess potential antioxidative capacity [21]. For instance, finger millet grain is found to have a dark brown coat that is rich in polyphenols, mainly ferulic acid, p-coumaric acid and proto-catechuic acid. They are also found to contain gallic acid, p-hydroxy benzoic acid, syringic acid and trans-cinnamic acid, which are derivatives of benzoic acid and cinnamic acid.

Ofosu et al. screened the different types of millets, including Italian millet, barnyard millet and finger Italian millet, to study the effect of various polyphenolic compounds that have antioxidant and antidiabetic activity and found that the food grains are rich in phenolic compounds such as catechin and its derivatives, dihydromyricetin, isorhamnetin, epicatechin, dihydroquercetin, dimethyl quercetin, formononetin, luteolin, kaempferol and apigenin [38].

In a similar research, Korean millet varieties were screened for their antioxidant and antidiabetic potential, and the researches showed that the finger Italian millets have the highest content of phenolics, flavonoids and condensed tannins, and the barnyard and finger Italian millets expressed the highest antioxidant capacity, which show their potential role in the prevention and management of chronic diseases such as type 2 diabetes, cancer and cardiovascular diseases [39].

9.10 PHENOLIC ACID-BOUND ARABINOXYLAN

Studies by Bijalwan et al. have shown that hydroxy-cinnamic acid-bound arabinoxylans extracted from the Indian millet grains, finger, proso, barnyard and foxtail millet expresses antioxidant potential. Similarly, fox tail and barnyard millet have shown immunostimulatory activity owing to their bound phenolic acids (mainly caffeic acid, p-coumaric acid and ferulic acid) with arabinoxylan. Rao et al. have shown that the feraxans, which is a water-soluble feruloyl arabinoxylans, present in finger millets exhibits approximately 5,000-fold higher antioxidative properties in comparison to the sulphated polysaccharides [40,41]. Srinivasan et al. report immunostimulatory activities of phenolic acid-bound arabinoxylans extracted from the foxtail and barnyard millets [42].

9.11 FLAVONOIDS AND NON-FLAVONOIDS

Millets are rich sources of flavonoids, especially flavones C-glycosides. Watanabe et al. showed that Japanese barnyard millet has high concentrations of strong antioxidants *N*-(p-coumaroyl)-serotonin and the flavones luteolin and tricin [43]. McDonough et al. studied the finger millet using scanning electron and fluorescent microscopy and had shown that the finger millets are rich in phenols and tannins [44,45]. In addition to the polyphenols, finger millet seed coat is also enriched with the flavonoid quercetin and phenolic compounds, such as naringenin, kaempferol, luteolin glycoside, phloroglucinol, apigenin, catechin/epicatechin, epigallocatechin, daidzein and catechin gallates [46–53]. Non-flavonoids comprising a single aromatic ring includes xanthones, stilbenes, lignans and tannins [54].

9.11.1 *Setaria italic* (Foxtail Millet)

Foxtail millet (Figure 9.1) is one of the most common millet varieties cultivated in Asian countries that are found to be associated with immense nutritional and health benefits. Zhang et al. found that the dietary intake of foxtail millet reduces the inflammation and risk of azoxymethane/dextran sodium sulphate-induced colitis-associated colorectal cancer in mice [55]. Kuruburu et al. investigated the effect of phenolic compounds and flavonoids in the fractions of foxtail millet on breast cancer cell lines and noticed the retardation of the cell proliferation by inducing G2/M cell cycle arrest and fragmentation of DNA [56].

Shan et al. isolated a novel protein called foxtail millet bran protein (FMBP) that is pharmaceutically very active as an anticancer agent as well as an immunomodulatory, antimicrobial and antidiabetic agent. The protein inhibits cell proliferation in colon cancer cells by mediating G1 cell cycle arrest and caspase-dependent apoptosis. Also, they inhibited the xenografted tumour growth in the nude mice model [57]. Composite nanoparticles comprising foxtail millet prolamin and curcumin-loaded chondroitin sulphate with sodium caseinate resulted in enhanced inhibition of tumour growth against HepG2 hepatoma cells. Several researches have been made on bound polyphenols of the inner shells (BPIS) of foxtail millet for their antioxidant activity and their ability to inhibit proliferation of cancer cells [58].

FIGURE 9.1 Small grained cereal foxtail millet (*Setaria italica*) found to contain phenolic compounds, flavonoids and anticancer bran proteins prevent proliferation of cancer cells and inhibition of tumor growth.

Earlier studies by Shi have shown that BPIS inhibits the proliferation through induction of apoptosis in human colorectal cancer HCT-116 cell and in HCT-116-bearing nude mice [59]. In a similar study, BPIS is evaluated in the azoxymethane and dextran sodium sulphate-induced mouse colorectal cancer model that showed reduction in the number and volume of the tumour models by protecting the intestinal epithelial cells from damage [60]. Zhang et al. have shown that the syringic acid and vitexin of the foxtail millet bran suppress the proliferation of the breast cancer cell lines through glucose-regulated protein 78 and sterol regulatory element-binding protein (GRP78/SREBP-1/SCD1) signalling [61].

Lu et al. investigate BPIS of foxtail millets and have found that 12 major phenolic compounds, especially ferulic acid and p-coumaric acid, enhanced the sensitivity of chemotherapy and apoptotic cell death in HCT-8/Fu colorectal cancer cells, thereby overcoming multidrug resistance [62]. Further studies have shown that the antimultidrug resistance is due to increase of miR-149 expression induced by methylation inhibition in the HCT-8/Fu colorectal cancer cells; and the cell surface glucose-regulated protein 78 (csGRP78) expressed on colorectal cancer cells could be associated with the anticancer effect of the colorectal cancer cells by the foxtail millet bran proteins [63,64].

9.11.2 *Panicum miliaceum* (Proso/White Millet)

When compared to the other common millets, proso millets (Figure 9.2) are very much considered as functional foods because of their potent antioxidative and

FIGURE 9.2 Proso millet (*Panicum miliaceum*) having effective inhibition of cancer cells due to peroxidases that protects oxidative damage in cells and a high concentration of phenolic compounds in the bound as well as free fractions.

antiproliferative activities. When compared to the other phenolic compounds, ferulic acid and chlorogenic acid are the predominant ones in the bound fraction of millets, although caffeic acid and p-coumaric acid are also present. And, syringic acid is present in the proso millet's free fraction. The proso extract exhibited a strong antioxidant potential and inhibited the proliferation of MDA breast cancer and HepG2 human hepatoma cell line [65,66]. Han et al. studied the inhibition of cell proliferation mediated by proso millet on MDA-MB-231 breast cancer cell lines and have shown that the apoptotic pathway is mediated through the activation of the BAK and BAX pathways and the activation of caspase cascades mediated through mitochondrial damage [67].

Cui et al. extracted the enzyme cationic peroxidase from proso millet that exhibited necroptosis in HCT116 and HT29 human colon cancer cell line through autocrine production of TNF-alpha and demethylation of the RIPK3 demethylation. Further research also shows that the double-stranded DNA breaks mediated by the enzyme were mainly responsible for the DNA damage in the colorectal carcinoma cells [68,69].

Ramadoss and Sivalingam isolated the phenolic aldehyde vanillin from both the proso and barnyard millet that inhibited the in vitro cell proliferation of HT-29 colon cancer cell line and human breast cancer cell line MCF-7 through apoptosis mediated by cell cycle inhibition [70–72]. In a similar research, Aburai et al. isolated linoleic acid from the proso and the Japanese millet that showed potent inhibition of the enzyme histone deacetylase and showed inhibition of human leukaemia K562 and prostate cancer LNCaP cells [73].

9.11.3 *Urochloa ramose* (Browntop Millet) and *Pennisetum glaucum* (Pearl Millet)

Sunagar and Sreerama used various solvents differing in their polarity to extract phenolic compounds from brown top millet (Figure 9.3) and also studied the antioxidant and enzyme inhibitory potential of the phenolic compounds. They found that the solvents acetone and ethanol were very efficient in extracting the phenolic compounds, particularly gallic acid, ferulic acid and caffeic acid. The flavonoids myricetin and kaempferol are also available and together the phenolics and the flavonoids are contributing to the strong antioxidant activity and preventing the oxidative damage associated with cancer. Besides, they also are found to inhibit the enzymes alpha-glucosidase and alpha-amylase that help in the digestion of carbohydrates and rapidly increase the blood glucose levels [74]. Pearl millet (Figure 9.4) also comprises bioactive phenolic compounds very effective as antioxidant agents for the prevention and treatment of cancers [75].

9.11.4 *Eleusine coracana* (Finger Millet)

Kuruburu studied the free and bound phenolic compounds of finger millet (Figure 9.5) and found that procyanidin B1, 3-O-methyl quercetin and epigallocatechin are the predominant free phenolic compounds; and protocatecheuic acid, ferulic acid and vanillic acid are the main bound phenolic compounds. When tested against the breast cancer cell lines, the free phenolic compounds exhibited an enhanced cytotoxicity in

FIGURE 9.3 Browntop millet (*Urochloa ramose*) rich in bioactive phenolic acids like gallic, ferulic and caffeic acid, and also flavonoids myricetin and kaempferol contributing to their anticancer activity.

FIGURE 9.4 Pearl millet (*Pennisetum glaucum*) is a promising cereal crop not only pertaining to its exceptional resistance to draught but also its effectiveness in cancer prevention and treatment

FIGURE 9.5 Finger millet (*Eleusine coracana*) is a highly valued small grain rich in free and bound phenolics and other anticancer compound like tannins rendering them effective against cancers and other oxidative stress related disorders.

comparison to the bound phenolic compounds by inducing DNA fragmentation and promoting apoptosis [76].

Samarasekera et al. analysed the Srilankan finger millet varieties for their anticancer potential and found that the alcoholic extracts are active against the human liver and breast cancer cell lines by inhibiting glutathione S transferase and beta-glucuronidase enzymes [77]. The predominant polyphenols of the finger millets are gallic, protocatechuic, p-hydroxybenzoic, vanillic and ferulic acids; syringic, trans-cinnamic and p-coumaric acids; and the flavonoid quercetin [78,79]. In a similar study, Siwela et al. analysed 22 types of finger millets grown in Southern and East Africa for the presence of tannins and their location and also their antioxidant properties and found that there are a wide range of phenolic compounds and condensed tannins with potential antioxidant activity, thereby rendering them an effective functional food against cancers and other oxidative stress-related conditions [80].

9.11.5 *Panicum sumatrense* (Little Millet)

Little millet (Figure 9.6) is also one of the common food grains desired for its high nutritional content and health benefits [81,82]. Besides being a nutritional food, little millet is also considered as the functional food due to the presence of pharmaceutically active components such as phenols and flavonoids that are effective in the prevention and treatment of oxidative stress-related diseases and disorders such as cancers, inflammatory diseases, diabetes and cardiovascular conditions [83–86].

FIGURE 9.6 Little millet (*Panicum sumatrense*) well known for promoting digestive health owing to the high fiber content also is rich in polyphenolic compounds and flavonoids contributes to lowering the risk of cancer development and treatment of oxidative stress related conditions.

Pradeep et al. compared six cultivars of foxtail and little millet for their phenolic contents and found that little millet has a higher phenolic content than that of the foxtail millet. The predominant phenolic compounds are ferulic, caffeic and sinapic acids, and the flavonoids were luteolin, quercetin, daidzein and kaempferol. The bioactive compounds exhibited antioxidant activities and the little millets exhibited a higher inhibition of α-amylase and α-glucosidase enzymes than foxtail millets, suggesting their effectiveness in the prevention and management of hyperglycaemia and cancers [87,88].

Similarly, Chandrasekara and Shahidi studied the antiproliferative effect of the phenolic compounds and flavonoid contents from seven millet varieties, including the little millet, and found that the inhibition of cell proliferation under in vitro condition and inhibition of peroxyl radical-induced DNA scission were very high, indicating the potential of the millet grains in the prevention of initiation and progression of cancer [27].

9.11.6 *Echinochloa spp.* (Barnyard Millet)

Barynard millet (Figure 9.7) is well known for its excellent nutritional profile, especially due to its richness in protein, carbohydrate, micronutrient and fibre contents and is generally used as traditional food [89–93]. Besides the high nutrient content and dietary value, they are found to contain several pharmaceutically active agents effective against disorders, particularly oxidative stress disorders, diabetes and cancers. Ofosu et al. have shown that barnyard millets comprise high amounts of

FIGURE 9.7 Barnyard millet (*Echinochloa spp.*) is another cereal crop that protects oxidative DNA damage and are rich sources of numerous anticancer bioactive compounds playing crucial role in the cellular processes associated to the prevention and treatment of cancers.

polyphenolic antioxidants, namely (+)-catechin, isorhamnetin, 3,7-dimethyl quercetin, formononetin, luteolin, kaempferol and apigenin.

Anis and Sreerama have evaluated the effectiveness of barnyard millet in controlling the protein glycation process that plays a key role in the prevention of diabetic complications, and their studies have shown that the major phenolic acids of the barnyard millet p-coumaric acid and chlorogenic acids exhibited the potent anti-glycation activity. Also, these phenolic acids play an essential role as antioxidants, as they were found to protect the oxidative DNA damage and also the process of hydroxy radical-induced protein fragmentation [94,95].

Similarly, Kim et al. screened 13 varieties of barnyard millets of Korea and found that all of them exhibited high quantities of total phenolic and flavonoid bioactive compounds, making them a potential functional food used for the prevention and treatment of cancers [96]. The Japanese barnyard millet was studied by Watanabe et al. and was found to contain *N*-(p-coumaroyl)-serotonin, luteolin and tricin, which are potential anticancer bioactive compounds [43].

Scientists have evaluated the anticancer potential of barnyard millets in various in vitro cell culture models and found the mediation of several cellular processes associated with the prevention and treatment of cancers. Kim analysed the flavonoid kaempferol isolated from the barnyard millet, which are effective in inducing G2 checkpoint arrest in the cell cycle, resulting in mitochondria-dependent apoptosis in human acute leukaemia Jurkat T cells [96]. Studies by Ramadoss and Sivalingam suggest that the phenolic aldehyde vanillin extracted from the proso and barnyard millets inhibited the cell proliferation of HT-29 colon cancer cell line, human breast cancer cell line MCF-7 and let to the apoptotic pathway by inducing cell cycle inhibition [70–72].

Barnyard millets are also well known for their richness in some of the bioactive antioxidants such as beta-glucan, gamma amino butyric acid and hydroxy-cinnamic acid-bound arabinoxylans [40,97], which are associated with antimutagenic effects by reducing the DNA damage, thereby slowing down the growth of cancer cells and also inhibiting the migration of cancer cells. Lee studied the barnyard millet for their anti-inflammatory activity and found that kaempferol and the isoflavone biochanin-A were mainly responsible for inhibiting the Lipopolysaccharide (**LPS**) -induced NO production in mouse macrophage cell lines. However, the phenolic compounds such as formononetin, hesperidin, naringin and protocatechuic acid also support this bioactivity [98]. In a similar study, Hosoda et al. evaluated the bran extracts of Japanese barnyard millet for their anti-inflammatory activity and found that the phenolics of the millet suppressed the production of NO and the cytokines associated with inflammation in mouse macrophage cell lines [99].

9.11.7 *Sorghum spp.* (Great Millet or Sorghum)

Great millets (Figure 9.8) are rich in phytochemicals such as phenolic acids (especially ferulic acid and gallic acid), flavonoids (mainly luteolin and apigenin), tannins, stilbenes, phytosterols, polycosanols and phenolamides and are effective against several health diseases and disorders associated with the accumulation of

FIGURE 9.8 Great millet (*Sorghum bicolor*) possesses potent antiproliferative activity due to the presence of a wide range of antioxidants such as phenolic acids, flavonoids, tannins, stilbenes and phytosterols.

reactive oxygen species such as inflammatory diseases, obesity and cancers and also age-related diseases like arthritis, diabetes, ageing, high cholesterol disorders and neurodegenerative diseases [45,100–103].

Especially, sorghum is found to exhibit strong antiproliferative activity when tested against several in vitro cancer cell lines, thereby acting as a potential therapeutic agent in preventing and inhibiting tumour growth. Lee et al. studied the cytotoxic potential of sorghum bran extract rich in phenolic compounds against human colon cancer cells. The results showed that the bran extract induces apoptosis and growth arrest in the S phase, stimulates DNA damage by altering the expression of metastasis-regulating genes and protects against human colon cancer [104]. Zhu et al. observed antiproliferative and cell migration inhibitory activity of proanthocyanidin-rich sorghum bran extracts against hepatocellular carcinoma cells [105].

Studies by Park et al. using Hwanggeumchal sorghum extract showed the suppression of tumour growth by modulating the Jak/STAT pathways on MDA-MB 231 and MCF-7 breast cancer animal models. In addition to inhibiting the STAT5b/IGF-1R and STAT3/VEGF pathways, the extract also was found to be associated with the downregulation of angiogenic factors and cell cycle regulators [106]. Also, the extract was found to promote bone growth through the activation of STAT-5B pathway evident with induced BMP7 and GH signalling in MC3T3-E1 osteoblastic cells [107].

They are rich in 3-deoxyanthoxyanins that have the potential to induce the activity of phase II enzymes in murine hepatoma cells and inhibit the growth of HT-29 human colon cancer cells [108]. Similarly, Suganyadevi et al. studied the 3-deoxyanthocyanins extracted from the bran of red sorghum and found that the phytochemical exhibited 84.09% inhibition of the MCF-7 human breast cancer cell line by mediating apoptosis through the upregulation of the p53 proapoptotic gene expression and downregulation of the Bcl-2 antiapoptotic gene expression [109]. Owumi et al. investigated the ability of three apigeninidin-rich extracts of sorghum for their cytotoxic activity against cancers, especially lung adenocarcinoma, and found that they inhibited oxidative and nitrosative stress and inflammation in the liver and kidney of rats treated with aflatoxin B1 [110]. Especially, red sorghum extracts are rich in phenolic compounds and the two 3-deoxyanthocyanidins luteolinidin and apigeninidin that exhibit potential antioxidant activities [101].

9.11.8 *Paspalum scrobiculatum* (Kodo Millet)

Several researches have been done on the antioxidant properties of the kodo millet (Figure 9.9), and the bioactive compounds, particularly the polyphenolic compounds and their derivatives, are mainly responsible for the antiproliferative and antiradical properties of the millets [111,112]. Sharma et al. have shown enhanced in vitro antioxidant potential of kodo millets that are processed and nutritionally enriched for better nutritional and nutraceutical properties such as enhanced total antioxidant capacity, metal chelating ability and hydrogen peroxide scavenging activities [113,114].

FIGURE 9.9 Kodo millet (*Paspalum scrobiculatum*) has enhanced anti-radical and antiproliferative properties owing to the presence of polyphenolic compounds and phenolic bound arabinoxylans providing health benefits against oxidative stress conditions

The polyphenolic extract of kodo millet was shown to be rich in phenolic acids such as quercetin, ferulic acid and cinnamic acid. Ramasamy et al. tested the polyphenolic extract of the kodo millet and observed that they have the potential to induce apoptosis in human cervical cancer cell lines HeLa. They observed growth inhibition evident with the signs of apoptosis such as membrane blebbing, appearance of apoptotic bodies, loss of mitochondrial membrane potential, increased production of reactive oxygen species and DNA damage. The apoptotic process was also found to be associated with the activation of caspase-9 and Poly (**ADP**-ribose) **polymerases** (PARPs) cleavage [115].

Srinivasan et al. studied the phenolic-bound arabinoxylans extracted from the little and kodo millet seeds and found that they have immunomodulatory effect on RAW 264.7 macrophage cell lines through the activation of Nuclear factor kappa B (**NF-κB**) and Extracellular signal-regulated kinase (ERK) pathways [116]. The results have shown that the kodo millet, in addition to providing the other health benefits associated with oxidative stress, are also a potential immunostimulatory agent that offers resistance to infections and diseases.

9.12 CONCLUSION

Millets are one of the staple food crops cultivated by the indigenous population of the world throughout the history; however, recently, they have been less appreciated. Understanding the significance of cultivating this neglected and underutilized crop and increasing their agricultural productivity will eventually lead to increased food production and reduces the dependability on major food crops such as rice, wheat and maize that have limited nutritional benefits. Millets are a nutrient-rich food that contribute to food diversity by offering healthy and diverse diets. They are rich in several pharmaceutically active phytochemicals such as polyphenols, which are potential and safe alternatives to medicines for the prevention, treatment and management of chronic diseases and disorders like cancer, obesity, inflammatory conditions, cardiovascular diseases and diabetes.

One of the Sustainable Development Goals of United Nations is to combat hunger, reduce malnutrition and promote the well-being of the world population by 2030. Understanding the significance and importance of millets in terms of their agricultural productivity, resistance to biotic and abiotic stress, health and nutritional benefits and their immense potential to reduce the risk of chronic diseases and disorders will certify millet as a functional food that has nutritional benefits as well as offer additional health benefits. Consequently, millets can be considered as an economically viable and a climate-resilient staple food crop and will certainly have a holistic impact in the global efforts ensuring food security and nutrition in the course of development of a better and sustainable future.

REFERENCES

1. Liu, X., Jones, M. K., Zhao, Z., Liu, G., & O'Connell, T. C. (2012 Oct). The earliest evidence of millet as a staple crop: new light on neolithic foodways in North China. *American Journal of Biological Anthropology*, 149(2), 283–290. https://doi.org/10.1002/ajpa.22127

2. Adekunle, A. A., Ellis-Jones, J., Ajibefun, I., Nyikal, R. A., Bangali, S., Fatunbi, A. O., & Angé, A. (2013). *Agricultural innovation in sub-Saharan Africa: experiences from multiple stakeholder approaches*. Accra, Ghana: Forum for Agricultural Research in Africa (FARA).
3. Ghatak, A., Chaturvedi, P., Bachmann, G., Valledor, L., Ramšak, Ž., Bazargani, M. M., Bajaj, P., Jegadeesan, S., Li, W., Sun, X., Gruden, K., Varshney, R. K., & Weckwerth, W. (2021). Physiological and proteomic signatures reveal mechanisms of superior drought resilience in pearl millet compared to wheat. *Frontiers in Plant Science*, 11, 600278. https://doi.org/10.3389/fpls.2020.600278
4. Ji, Y., Lu, X., Zhang, H., Luo, D., Zhang, A., Sun, M., Wu, Q., Wang, X., & Huang, L. (2021). Transcriptome reveals the dynamic response mechanism of pearl millet roots under drought stress. *Genes*, 12(12), 1988. https://doi.org/10.3390/genes12121988
5. Chandrashekar, A., & Satyanarayana, K. V. (2006). Disease and pest resistance in grains of sorghum and millets. *Journal of Cereal Science*, 44(3), 287–304, https://doi.org/10.1016/j.jcs.2006.08.010
6. Sanogo, M. D., & de Milliano, W. A. (2007). Accumulation of pearl millet downy mildew resistance in Mali–2006 results. *Communications in Agricultural and Applied Biological Sciences*, 72(2), 303–308.
7. Manjunatha, G., Roopa, K. S., Prashanth, G. N., & Shetty, H. S. (2008). Chitosan enhances disease resistance in pearl millet against downy mildew caused by Sclerospora graminicola and defence-related enzyme activation. *Pest Management Science*, 64(12), 1250–1257. https://doi.org/10.1002/ps.1626
8. Williams, R. J., & Andrews, D. J. (1983). Breeding for disease and pest resistance in pearl millet. In: *FAO/IITA expert consultation on durable resistance breeding*, 23–29 October 1982, Ibadan, Nigeria.
9. Padulosi, S., Thompson, J., & Rudebjer, P. G. (2013). *Fighting poverty, hunger and malnutrition with neglected and underutilized species: needs, challenges and the way forward.* Journal of Ethnic Foods, 8(1), Article 5. https://doi.org/10.1186/s42779-021-00030-0
10. Ravi, S. B., Hrideek, T. K., Kumar, A. T., Prabhakaran, T. R., Mal, B., & Padulosi, S. (2010). Mobilizing neglected and underutilized crops to strengthen food security and alleviate poverty in India. *Indian Journal of Plant Genetic Resources*, 23(1), 110–116.
11. Chandra, M. S., Naresh, R. K., Thenua, O. V. S., Singh, R., & Geethanjali, D. (2020). Improving resource conservation, productivity and profitability of neglected and underutilized crops in the breadbasket of India: a review. *Pharma Innovation*, 9(3), 685–696.
12. Li, X., Yadav, R., & Siddique, K. H. M. (2020). Neglected and underutilized crop species: the key to improving dietary diversity and fighting hunger and malnutrition in Asia and the Pacific. *Frontiers in Nutrition*, 7, 593711. https://doi.org/10.3389/fnut.2020.593711
13. Hunter, D., Borelli, T., Beltrame, D. M. O., Oliveira, C. N. S., Coradin, L., Wasike, V. W., Wasilwa, L., Mwai, J., Manjella, A., Samarasinghe, G. W. L., Madhujith, T., Nadeeshani, H. V. H., Tan, A., Ay, S. T., Güzelsoy, N., Lauridsen, N., Gee, E., & Tartanac, F. (2019). The potential of neglected and underutilized species for improving diets and nutrition. *Planta*, 250(3), 709–729. https://doi.org/10.1007/s00425-019-03169-4
14. Rawat, L., Karnatak, A. K., Bisht, T. S., & Kukreti, A. (2021). Minor millets: profile and ethnobotanical scenario. *Millets and Millet Technology*. Springer, Singapore, 51–80. https://doi.org/10.1007/978-981-16-0676-2_3
15. Garg, A., Joshi, K., & Raghav, P. K. (2022). Smart food for better health and nutrition–a promising concept. *Journal of Postharvest Technology*, 10(4), 179–190.
16. Weerasekera, A. C., Samarasinghe, K., & Waisundara, V. Y. (2022). Introductory chapter: nutritive value of pseudocereals. *Pseudocereals*, 1. IntechOpen. Available at: http://dx.doi.org/10.5772/intechopen.101720
17. Shahidi, F., & Chandrasekara, A. (2013). Millet grain phenolics and their role in disease risk reduction and health promotion: a review. *Journal of Functional Foods*, 5(2), 570–581.

18. Mohamed, T. K., Issoufou, A., & Zhou, H. (2012). Antioxidant activity of fractionated foxtail millet protein hydrolysate. *International Food Research Journal*, 19(1), 93–100.
19. Kalinova, J., & Moudry, J. (2006). Content and quality of protein in proso millet (Panicum miliaceum L.) varieties. *Plant Foods for Human Nutrition*, 61(1), 43–47. https://doi.org/10.1007/s11130-006-0013-9
20. Amadou, I., Gounga, M. E., & Le, G. W. (2013). Millets: nutritional composition, some health benefits and processing – a review. *Emirates Journal of Food and Agriculture*, 25(7), 501–508. https://doi.org/10.1094/CFW-56-3-0115
21. Chandrasekara, A., Joshepkumar, T., Sakthivel, M., & Marian-Victor, V. (2013). Phenolic content and antioxidant activity of raw and processed roots and tuber crops. *Annals of Nutrition and Metabolism*, 63, 1594.
22. Shi, H., Shi, G. Y., Yang, C. Y., Shi, G. S., Hou, G. L., Zhang, C. P., & Cheng, Y. (2007). Study on the nourishment health care of millet and food treat value. *Rain Fed Crops*, 27, 376–378.
23. Huo, Q. G., Bi, Y. L., Zhu, X. P., & Wang, M. M. (2006). Study on components of millet oil. *Journal of China Oil*, 31, 63–64.
24. Liang, S., & Liang, K. (2019). Millet grain as a candidate antioxidant food resource: a review. *International Journal of Food Properties*, 22(1), 1652–1661, https://doi.org/10.1080/10942912.2019.1668406
25. Hassan, Z. M., Sebola, N. A., & Mabelebele, M. (2021). The nutritional use of millet grain for food and feed: a review. *Agriculture & Food Security*, 10(1), 16. https://doi.org/10.1186/s40066-020-00282-6
26. Majid, A., & Priyadarshini, C. G. P. (2020). Millet derived bioactive peptides: a review on their functional properties and health benefits. *Critical Reviews in Food Science and Nutrition*, 60(19), 3342–3351. https://doi.org/10.1080/10408398.2019.1686342
27. Chandrasekara, A., & Shahidi, F. (2010). Content of insoluble bound phenolics in millets and their contribution to antioxidant capacity. *Journal of Agricultural and Food Chemistry*, 58(11), 6706–6714. https://doi.org/10.1021/jf100868b
28. Dhliwayo, T., Nyanga, L. K., Chopera, P., Matsungo, T. M., & Chidewe, C. (2022). A review of traditional grain-based complementary foods for children aged 6–23 months in selected African countries. *The North African Journal of Food and Nutrition Research*, 6(14), 115–125.
29. Zhu, F. (2019). Proanthocyanidins in cereals and pseudocereals. *Critical Reviews in Food Science and Nutrition*, 59(10), 1521–1533. https://doi.org/10.1080/10408398.2017.1418284
30. Hung, P. V. (2016). Phenolic compounds of cereals and their antioxidant capacity. *Critical Reviews in Food Science and Nutrition*, 56(1), 25–35. https://doi.org/10.1080/10408398.2012.708909
31. Zhang, B., Liu, J., Cheng, L., Zhang, Y., Hou, S., Sun, Z., Li, H., & Han, Y. (2019). Carotenoid composition and expression of biosynthetic genes in yellow and white foxtail millet [Setaria italica (L.) Beauv]. *Journal of Cereal Science*, 85, 84–90, https://doi.org/10.1016/j.jcs.2018.11.005.
32. Amadou, I., Le, G. W., & Shi, Y. H. (2013). Evaluation of antimicrobial, antioxidant activities, and nutritional values of fermented foxtail millet extracts by *Lactobacillus paracasei* Fn032. *International Journal of Food Properties*, 16(6), 1179–1190.
33. Fernandez-Orozco, R., Frias, J., Zielinski, H., Piskula, M. K., Kozlowska, H., & Vidal-Valverde, C. (2008). Kinetic study of the antioxidant compounds and antioxidant capacity during germination of Vigna radiata cv. emmerald, Glycine max cv. jutro and Glycine max cv. merit. *Food Chemistry*, 111(3), 622–630.

34. Banwo, K., Asogwa, F. C., Ogunremi, O. R., Adesulu-dahunsi, A., & Sanni, A. (2021). Nutritional profile and antioxidant capacities of fermented millet and sorghum gruels using lactic acid bacteria and yeasts. *Food Biotechnology*, 35(3), 199–220. https://doi.org/10.1080/08905436.2021.1940197
35. Lü, J. M., Lin, P. H., Yao, Q., & Chen, C. (2010 Apr). Chemical and molecular mechanisms of antioxidants: experimental approaches and model systems. *Journal of Cellular and Molecular Medicine*, 14(4), 840–860. https://doi.org/10.1111/j.1582-4934.2009.00897.x
36. Panchatcharam, M., Miriyala, S., Gayathri, V. S., & Suguna, L. (2006). Curcumin improves wound healing by modulating collagen and decreasing reactive oxygen species. *Molecular and Cellular Biochemistry*, 290, 87–96.
37. Shih, P. H., Yeh, C. T., & Yen, G. C. (2007). Anthocyanins induce the activation of phase II enzymes through the antioxidant response element pathway against oxidative stress-induced apoptosis. *Journal of Agricultural and Food Chemistry*, 55, 9427–9435.
38. Ofosu, F. K., Elahi, F., Daliri, E. B., Chelliah, R., Ham, H. J., Kim, J. H., Han, S. I., Hur, J. H., & Oh, D. H. (2020). Phenolic profile, antioxidant, and antidiabetic potential exerted by millet grain varieties. *Antioxidants*, 9(3), 254. https://doi.org/10.3390/antiox9030254
39. Kim, J. Y., Jang, K. C., Park, B. R., Han, S. I., Choi, K. J., Kim, S. Y., Oh, S. H., Ra, J. E., Ha, T. J., Lee, J. H., & Hwang, J. (2011). Physicochemical and antioxidative properties of selected barnyard millet (Echinochloa utilis) species in Korea. *Food Science and Biotechnology*, 20, 461–469.
40. Bijalwan, V., Ali, U., Kesarwani, A. K., Yadav, K., & Mazumder, K. (2016). Hydroxycinnamic acid bound arabinoxylans from millet brans-structural features and antioxidant activity. *International Journal of Biological Macromolecules*, 88, 296–305. https://doi.org/10.1016/j.ijbiomac.2016.03.069
41. Rao, R. S. P., & Muralikrishna, G. (2006). Water soluble feruloyl arabinoxylans from rice and ragi: changes upon malting and their consequence on antioxidant activity. *Phytochemistry*, 67(1), 91–99.
42. Srinivasan, A., Ekambaram, S. P., Perumal, S. S., Aruldhas, J., & Erusappan, T. (2020). Chemical characterization and immunostimulatory activity of phenolic acid bound arabinoxylans derived from foxtail and barnyard millets. *Journal of Food Biochemistry*, 44(2), e13116. https://doi.org/10.1111/jfbc.13116
43. Watanabe, M. (1999). Antioxidative phenolic compounds from Japanese barnyard millet (Echinochloa utilis) grains. *Journal of Agricultural and Food Chemistry*, 47(11), 4500–4505. https://doi.org/10.1021/jf990498s
44. McDonough, C. M., Rooney, L. W., & Earp, C. F. (1986). Structural characteristics of *Eleusine corocana* (finger millet) using scanning electron and fluorescence microscopy. *Food Structure*, 5(2), Article 9.
45. Dykes, L, & Rooney, L. M. (2006). Sorghum and millet phenols and antioxidants. *Journal of Cereal Science*, 44, 236–251.
46. Viswanath, V., Urooj, A., & Malleshi, N. G. (2009). Evaluation of antioxidant and antimicrobial properties of finger millet polyphenols (Eleusine coracana). *Food Chemistry*, 114, 340–346.
47. Devi, P. B., Vijayabharathi, R., Sathyabama, S., Malleshi, N. G., & Priyadarisini, V. B. (2014). Health benefits of finger millet (Eleusine coracana L.) polyphenols and dietary fiber: a review. *Journal of Food Science and Technology*, 51, 1021–1040. https://doi.org/10.1007/s13197-011-0584-9
48. Shobana, S., Sreerama, Y. N., & Malleshi, N. G. (2009). Composition and enzyme inhibitory properties of finger millet (Eleusine coracana L.) seed coat phenolics: mode of inhibition of α-glucosidase and pancreatic amylase. *Food Chemistry*, 115, 1268–1273.

49. Rao, M. V. S. S. T. S., & Muralikrishna, G. (2002). Evaluation of the antioxidant properties of free and bound phenolic acids from native and malted finger millet (ragi, Eleusine coracana Indaf-15). *Journal of Agricultural and Food Chemistry*, 50, 889–892.
50. Chethan, S., & Malleshi, N. G. (2007). Finger millet polyphenols: optimization of extraction and the effect of pH on their stability. *Food Chemistry*, 105, 862–870.
51. Chethan, S., & Malleshi, N. G. (2007). Finger millet polyphenols: characterization and their nutraceutical potential. *American Journal of Food Technology*, 2, 582–592.
52. Chethan, S., Dharmesh, S. M., & Malleshi, N. G. (2008). Inhibition of aldose reductase from cataracted eye lenses by finger millet (Eleusine coracana) polyphenols. *Bioorganic & Medicinal Chemistry*, 16, 10085–10090.
53. Chethan, S., Sreerama, Y. N., & Malleshi, N. G. (2008). Mode of inhibition of finger millet malt amylases by the millet phenolics. *Food Chemistry*, 111, 187–191.
54. Siemińska-Kuczer, A., Szymańska-Chargot, M., & Zdunek, A. (2022). Recent advances in interactions between polyphenols and plant cell wall polysaccharides as studied using an adsorption technique. *Food Chemistry*, 373(Pt B), 131487. https://doi.org/10.1016/j.foodchem.2021.131487
55. Zhang, B., Xu, Y., Liu, S., Lv, H., Hu, Y., Wang, Y., Li, Z., Wang, J., Ji, X., Ma, H., Wang, X., & Wang, S. (2020). Dietary supplementation of foxtail millet ameliorates colitis-associated colorectal cancer in mice via activation of gut receptors and suppression of the STAT3 pathway. *Nutrients*, 12(8), 2367. https://doi.org/10.3390/nu12082367
56. Kuruburu, M. G., Bovilla, V. R., Leihang, Z., & Madhunapantula, S. V. (2022). Phytochemical-rich fractions from foxtail millet (Setaria italica (L.) P. Beauv) seeds exhibited antioxidant activity and reduced the viability of breast cancer cells in vitro by inducing DNA fragmentation and promoting cell cycle arrest. *Anti-Cancer Agents in Medicinal Chemistry*, 22(13), 2477–2493. https://doi.org/10.2174/1871520622666220215122141
57. Shan, S., Li, Z., Newton, I. P., Zhao, C., Li, Z., & Guo, M. (2014). A novel protein extracted from foxtail millet bran displays anti-carcinogenic effects in human colon cancer cells. *Toxicology Letters*, 227(2), 129–138. https://doi.org/10.1016/j.toxlet.2014.03.008
58. Chen, X., Wu, Y. C., Gong, P. X., Zhang, Y. H., & Li, H. J. (2022). Chondroitin sulfate deposited on foxtail millet prolamin/caseinate nanoparticles to improve physicochemical properties and enhance cancer therapeutic effects. *Food & Function*, 13(9), 5343–5352. https://doi.org/10.1039/d2fo00572g
59. Shi, J., Shan, S., Li, Z., Li, H., Li, X., & Li, Z. (2015). Bound polyphenol from foxtail millet bran induces apoptosis in HCT-116 cell through ROS generation. *Journal of Functional Foods*, 17, 958–968.
60. Yang, R., Shan, S., Zhang, C., Shi, J., Li, H., & Li, Z. (2020). Inhibitory effects of bound polyphenol from foxtail millet bran on colitis-associated carcinogenesis by the restoration of gut microbiota in a mice model. *Journal of Agricultural and Food Chemistry*, 68(11), 3506–3517. https://doi.org/10.1021/acs.jafc.0c00370
61. Zhang, L., La, X., Tian, J., Li, H., Li, A., Liu, Y., Wu, C., & Li, Z. (2021). The phytochemical vitexin and syringic acid derived from foxtail fillet bran inhibit breast cancer cells proliferation via GRP78/SREBP-1/SCD1 signaling axis. *Journal of Functional Foods*, 85, 104620.
62. Lu, Y., Shan, S., Li, H., Shi, J., Zhang, X., & Li, Z. (2018). Reversal effects of bound polyphenol from foxtail millet bran on multidrug resistance in human HCT-8/Fu colorectal cancer cell. *Journal of Agricultural and Food Chemistry*, 66(20), 5190–5199. https://doi.org/10.1021/acs.jafc.8b01659
63. Shan, S., Lu, Y., Zhang, X., Shi, J., Li, H., & Li, Z. (2021). Inhibitory effect of bound polyphenol from foxtail millet bran on miR-149 methylation increases the chemosensitivity of human colorectal cancer HCT-8/Fu cells. *Molecular and Cellular Biochemistry*, 476(2), 513–523. https://doi.org/10.1007/s11010-020-03906-4

64. Shan, S., Niu, J., Yin, R., Shi, J., Zhang, L., Wu, C., Li, H., & Li, Z. (2022). Peroxidase from foxtail millet bran exerts anti-colorectal cancer activity via targeting cell-surface GRP78 to inactivate STAT3 pathway. *Acta Pharmaceutica Sinica. B*, 12(3), 1254–1270. https://doi.org/10.1016/j.apsb.2021.10.004
65. Shen, R., Ma, Y., Jiang, L., Dong, J., Zhu, Y., & Ren, G. (2018). Chemical composition, antioxidant, and antiproliferative activities of nine Chinese proso millet varieties. *Food and Agricultural Immunology*, 29(1), 625–637. https://doi.org/10.1080/09540105.2018.1428283
66. Zhang, L., Liu, R., & Niu, W. (2014). Phytochemical and antiproliferative activity of proso millet. *PloS One*, 9(8), e104058. https://doi.org/10.1371/journal.pone.0104058
67. Han, C. R., & Kim, Y. H. (2023). Apoptotic effect of proso millet grains on human breast cancer MDA-MB-231 cells is exerted by activation of BAK and BAX, and mitochondrial damage-mediated caspase cascade activation. *J Life Science*, 33(1), 15–24.
68. Cui, X., Wang, R., & Wang, Z. (2018). Cationic peroxidase from proso millet induces human colon cancer cell necroptosis by regulating autocrine TNF-α and RIPK3 demethylation. *Food & Function*, 9(3), 1878–1888. https://doi.org/10.1039/c7fo01040k
69. Cui, X., Cao, Z., & Wang, Z. (2019). DNA damage and necroptosis induced by peroxidase from proso millet in human colorectal cancer cells. *Tropical Journal of Pharmaceutical Research*, 18(5), 975–983. https://dx. doi.org/10.4314/tjpr.v18i5.10
70. Ramadoss, D. P., & Sivalingam, N. (2017). Vanillin extracted from proso millet and barnyard millet induce apoptosis in HT-29 and MCF-7 cell line through mitochondria mediated pathway. *Asian Journal of Pharmaceutical and Clinical Research*, 10, 226–229.
71. Ramadoss, D. P., & Sivalingam, N. (2020). Vanillin extracted from proso and barnyard millets induce apoptotic cell death in HT-29 human colon cancer cell line. *Nutrition and Cancer*, 72(8), 1422–1437.
72. Ramadoss, D. P., & Sivalingam, N. (2021). Vanillin extracted from proso and barnyard millets induces cell cycle inhibition and apoptotic cell death in MCF-7 cell line. *Journal of Cancer Research and Therapeutics*, 17(6), 1425–1433.
73. Aburai, N., Esumi, Y., Koshino, H., Nishizawa, N., & Kimura, K. (2007). Inhibitory activity of linoleic acid isolated from proso and Japanese millet toward histone deacetylase. *Bioscience, Biotechnology, and Biochemistry*, 71(8), 2061–2064. https://doi.org/10.1271/bbb.70068
74. Sunagar, R. R., & Sreerama, Y. N. (2023). Implication of solvent polarities on browntop millet (Urochloa ramosa) phenolic antioxidants and their ability to protect oxidative DNA damage and inhibit α-amylase and α-glucosidase enzymes. *Food Chemistry*, 411, 135474.
75. Sharma, A., & Kapoor, A. C. (1996). Levels of antinutritional factors in pearl millet as affected by processing treatments and various types of fermentation. *Plant Foods Human Nutrition*, 1996, 49(3), 241–252.
76. Kuruburu, M. G., Bovilla, V. R., Naaz, R., Leihang, Z., & Madhunapantula, S. V. (2022). Variations in the anticancer activity of free and bound phenolics of finger millet (Eleusine coracana (L) Gaertn; variety KMR-301) seeds. *Phytomedicine Plus*, 2(2), 100276.
77. Samarasekera, J. K. R. R., Hettiarachchi, G. H. C. M., Gooneratne, M. J., Choudhary, M. I., Imad, R., & Naz, A. (2020). Glutathione S-transferase and β-glucuronidase enzymes inhibitory and cytotoxic activities of ethanolic and methanolic extracts of Sri Lankan finger millet (Eleusine coracana (L.) Gaertn.) varieties. *South Asian Research Journal of Natural Products*, 3(2), 1–9.
78. Banerjee, S., Sanjay, K. R., Chethan, S., & Malleshi, N. G. (2012). Finger millet (*Eleusine coracana*) polyphenols: investigation of their antioxidant capacity and antimicrobial activity. *African Journal of Food Science*, 6(13), 362–374.

79. Kumar, A., Rani, M., Mani, S., Shah, P., Singh, D. B., Kudapa, H., & Varshney, R. K. (2021). Nutritional significance and antioxidant-mediated antiaging effects of finger millet: molecular insights and prospects. *Frontiers in Sustainable Food Systems*, 5, 684318.
80. Siwela, M., Taylor, J. R., de Milliano, W. A., & Duodu, K. G. (2007). Occurrence and location of tannins in finger millet grain and antioxidant activity of different grain types. *Cereal Chemistry*, 84(2), 169–174.
81. Srilekha, K., Kamalaja, T., Uma Maheswari, K., & Neela Rani, R. (2019). Nutritional composition of little millet flour. *International Research Journal of Pure and Applied Chemistry*, 20(4), 1–4.
82. Saini, S., Saxena, S., Samtiya, M., Puniya, M., & Dhewa, T. (2021). Potential of underutilized millets as nutri-cereal: an overview. *Journal of Food Science and Technology*, 58(12), 4465–4477. https://doi.org/10.1007/s13197-021-04985-x
83. Kundgol, N. G., Kasturiba, B., Math, K. K., Kamatar, M. Y., & Usha, M. (2013). Impact of decortication on chemical composition, antioxidant content and antioxidant activity of little millet landraces. *International Journal of Advanced Research and Technology*, 2, 1705–1720.
84. Pujari, N., & Hoskeri, J. H. (2022). Minor millet phytochemicals and their pharmacological potentials. *Pharmacognosy Reviews*, 16(32), 101.
85. Jayalaxmi, R., Divya, K., Ankita, S., & Koteswaraiah, P. (2018). Antioxidant activities of pearl millet (Pennisetum glaucum) and little millet (Panicum sumatrense) in different in vitromodels. *Journal of Bioassays*, 7(2), 5595–5601.
86. Kumari, P., Singh, Y. V., Yadav, R., Vinay, K., Chandan, K. H., & Reddy, K. R. C. (2022). Analytical study of little millets and therapeutic effect of little millets based on its components in type 2 diabetes mellitus. *International Journal of Herbal Medicine*, 10, 11–14.
87. Pradeep, P. M., & Sreerama, Y. N. (2017). Soluble and bound phenolics of two different millet genera and their milled fractions: comparative evaluation of antioxidant properties and inhibitory effects on starch hydrolysing enzyme activities. *Journal of Functional Foods*, 35, 682–693. https://doi.org/10.1016/j.jff.2017.06.033
88. Pradeep, P. M., & Sreerama, Y. N. (2018). Phenolic antioxidants of foxtail and little millet cultivars and their inhibitory effects on α-amylase and α-glucosidase activities. *Food Chemistry*, 247, 46–55. https://doi.org/10.1016/j.foodchem.2017.11.103
89. Bhat, S., Nandini, C., & Tippeswamy, V. (2018). Significance of small millets in nutrition and health – a review. *Asian Journal of Dairy and Food Research*, 37(1), 35–40.
90. Bhatt, D., Rasane, P., Singh, J., Kaur, S., Fairos, M., Kaur, J., Fairos, M., Kaur, J., Gunjal, M., Mahato, D. K., Mehta, C., Avinashe, H., & Sharma, N. (2022). Nutritional advantages of barnyard millet and opportunities for its processing as value-added foods. *Journal of Food Science and Technology*, 1–13.
91. Renganathan, V. G., Vanniarajan, C., Karthikeyan, A., & Ramalingam, J. (2020). Barnyard millet for food and nutritional security: current status and future research direction. *Frontiers in Genetics*, 11, 500. https://doi.org/10.3389/fgene.2020.00500
92. Saleh, A. S., Zhang, Q., Chen, J., & Shen, Q. (2013). Millet grains: nutritional quality, processing, and potential health benefits. *Comprehensive Reviews in Food Science and Food Safety*, 12(3), 281–295.
93. Singh, A., Bharath, M., Kotiyal, A., Rana, L., & Rajpal, D. (2022). Barnyard millet: the underutilized nutraceutical minor millet crop. *Journal of Pharmaceutical Innovation*, 11(6), 115–128.
94. Anis, M. A., & Sreerama, Y. N. (2020). Inhibition of protein glycoxidation and advanced glycation end-product formation by barnyard millet (*Echinochloa frumentacea*) phenolics. *Food Chemistry*, 315, 126265. https://doi.org/10.1016/j.foodchem.2020.126265

95. Seo, K. H., Ra, J. E., Lee, S. J., Lee, J. H., Kim, S. R., Lee, J. H., & Seo, W. D. (2015). Anti-hyperglycemic activity of polyphenols isolated from barnyard millet (Echinochloa utilis L.) and their role inhibiting α-glucosidase. *Journal of the Korean Society for Applied Biological Chemistry*, 58, 571–579. https://doi.org/10.1007/s13765-015-0070-6
96. Kim, K. Y., Jang, W. Y., Lee, J. Y., Ko, J. Y., Yun, Y. H., & Kim, Y. H. (2016). Bioactive compounds/food microbiology: kaempferol activates G2-checkpoint of the cell cycle resulting in G2-arrest and mitochondria-dependent apoptosis in human acute leukemia Jurkat T cells. *Journal of Microbiology and Biotechnology*, 26(2), 287–294. https://doi.org/10.4014/jmb.1511.11054
97. Sharma, S., Saxena, D. C., & Riar, C. S. (2016). Isolation of functional components β-glucan and γ-amino butyric acid from raw and germinated barnyard millet (Echinochloa frumentaceae) and their characterization. *Plant Foods for Human Nutrition*, 71, 231–238. https://link.springer.com/article/10.1007/s11130-016-0545-6
98. Lee, J. Y., Jun, D. Y., Yoon, Y. H., Ko, J. Y., Woo, K. S., Woo, M. H., & Kim, Y. H. (2014). Anti-inflammatory effect of flavonoids kaempferol and biochanin a-enriched extract of barnyard millet (Echinochloa crus-galli var. frumentacea) grains in LPS-stimulated RAW264. 7 cells. *Journal of Life Science*, 24(11), 1157–1167.
99. Hosoda, A., Okai, Y., Kasahara, E., Inoue, M., Snhimizu, M., Usui, Y., Sekiyama, A., & Higashi-Okai, K. (2012). Potent immunomodulating effects of bran extracts of traditional Japanese millets on nitric oxide and cytokine production of macrophages (RAW264.7) induced by lipopolysaccharide. *Journal of UOEH*, 34(4), 285–296. https://doi.org/10.7888/juoeh.34.285
100. Xu, J., Wang, W., & Zhao, Y. (2021). Phenolic compounds in whole grain sorghum and their health benefits. *Foods (Basel, Switzerland)*, 10(8), 1921. https://doi.org/10.3390/foods10081921
101. Carbonneau, M. A., Cisse, M., Mora-Soumille, N., Dairi, S., Rosa, M., Michel, F., Lauret, C., Cristol, J. P., & Dangles, O. (2014). Antioxidant properties of 3-deoxyanthocyanidins and polyphenolic extracts from Côte d'Ivoire's red and white sorghums assessed by ORAC and in vitro LDL oxidisability tests. *Food Chemistry*, 145, 701–709. https://doi.org/10.1016/j.foodchem.2013.07.025
102. Li, Z., Zhao, X., Zhang, X., & Liu, H. (2021). Bioactive compounds and biological activities of sorghum grains. *Foods (Basel, Switzerland)*, 10(11), 2868. https://doi.org/10.3390/foods10112868
103. Rezaee, N., Fernando, W. M. A. D. B., Hone, E., Sohrabi, H. R., Johnson, S. K., Gunzburg, S., & Martins, R. N. (2021). Potential of sorghum polyphenols to prevent and treat Alzheimer's disease: a review article. *Frontiers in Aging Neuroscience*, 13, 729949. https://doi.org/10.3389/fnagi.2021.729949
104. Lee, S. H., Lee, J., Herald, T., Cox, S., Noronha, L., Perumal, R., Lee, H. S., & Smolensky, D. (2020). Anticancer activity of a novel high phenolic sorghum bran in human colon cancer cells. *Oxidative Medicine and Cellular Longevity*, e-collection 2020:2890536. https://doi.org/10.1155/2020/2890536
105. Zhu, Y., Shi, Z., Yao, Y., Hao, Y., & Ren, G. (2017). Antioxidant and anti-cancer activities of proanthocyanidins-rich extracts from three varieties of sorghum (*Sorghum bicolor*) bran. *Food and Agricultural Immunology*, 28(6), 1530–1543, https://doi.org/10.1080/09540105.2017.1351526
106. Park, J. H., Darvin, P., Lim, E. J., Joung, Y. H., Hong, D. Y., Park, E. U., Park, S. H., Choi, S. K., Moon, E. S., Cho, B. W., Park, K. D., Lee, H. K., Kim, M. J., Park, D. S., Chung, I. M., & Yang, Y. M. (2012). Hwanggeumchal sorghum induces cell cycle arrest, and suppresses tumor growth and metastasis through Jak2/STAT pathways in breast cancer xenografts. *PloS One*, 7(7), e40531. https://doi.org/10.1371/journal.pone.0040531

107. Joung, Y. H., Lim, E. J., Darvin, P., Jang, J. W., Park, K. D., Lee, H. K., Kim, H. S., Cho, B. W., Park, T., Chung, S., Park, J. H., & Yang, Y. M. (2013). Hwanggeumchal sorghum extract enhances BMP7 and GH signaling through the activation of Jak2/STAT5B in MC3T3-E1 osteoblastic cells. *Molecular Medicine Reports*, 8(3), 891–896. https://doi.org/10.3892/mmr.2013.1593
108. Yang, L., Browning, J. D., & Awika, J. M. (2009). Sorghum 3-deoxyanthocyanins possess strong phase II enzyme inducer activity and cancer cell growth inhibition properties. *Journal of Agricultural and Food Chemistry*, 57(5), 1797–1804. https://doi.org/10.1021/jf8035066
109. Suganyadevi, P., Saravanakumar, K. M., & Mohandas, S. (2013). The antiproliferative activity of 3-deoxyanthocyanins extracted from red sorghum (*Sorghum bicolor*) bran through P(53)-dependent and Bcl-2 gene expression in breast cancer cell line. *Life Sciences*, 92(6–7), 379–382. https://doi.org/10.1016/j.lfs.2013.01.006
110. Owumi, S. E., Kazeem, A. I., Wu, B., Ishokare, L. O., Arunsi, U. O., & Oyelere, A. K. (2022). Apigeninidin-rich Sorghum bicolor (L. Moench) extracts suppress A549 cells proliferation and ameliorate toxicity of aflatoxin B1-mediated liver and kidney derangement in rats. *Scientific Reports*, 12(1), 7438. https://doi.org/10.1038/s41598-022-10926-1
111. Dey, S., Saxena, A., Kumar, Y., Maity, T., & Tarafdar, A. (2022). Understanding the antinutritional factors and bioactive compounds of kodo millet (paspalum scrobiculatum) and little millet (panicum sumatrense). *Journal of Food Quality*, Vol. 2022, 1–19. DOI: 10.1155/2022/1578448.
112. Patil, R. B., Vijayalakshmi, K. G., & Vijayalakshmi, D. (2020). Physical, functional, nutritional, phytochemical and antioxidant properties of kodo millet (Paspalum scrobiculatum). *Journal of Pharmacognosy and Phytochemistry*, 9(5), 2390–2393.
113. Sharma, S., Jan, R., & Riar, C. S. (2021). Analyzing the effect of germination on the pasting, rheological, morphological and in-vitro antioxidant characteristics of kodo millet flour and extracts. *Food Chemistry*, 361, 130073.
114. Sharma, S., Sharma, N., Handa, S., & Pathania, S. (2017). Evaluation of health potential of nutritionally enriched kodo millet (Eleusine coracana) grown in Himachal Pradesh, India. *Food Chemistry*, 214, 162–168. https://doi.org/10.1016/j.foodchem.2016.07.086
115. Ramasamy, S., Anupam, K., Ratnavathi, C. V., Karunagaran, D., & Chandra, T. S. (2017). Paspalum scrobiculatum polyphenol rich extracts inhibit human cervical cancer cell proliferation and induce apoptosis in vitro. *The FASEB Journal*, 31, lb162–lb162.
116. Srinivasan, A., Aruldhas, J., Perumal, S. S., & Ekambaram, S. P. (2021). Phenolic acid bound arabinoxylans extracted from little and kodo millets modulate immune system mediators and pathways in RAW 264.7 cells. *Journal of Food Biochemistry*, 45(1), e13563. https://doi.org/10.1111/jfbc.13563

10 *Phyllanthus emblica* *A Promising Start in the Quest for a Cancer Drug*

Shalini Mukherjee and Md. Kausar Neyaz
Aabir Bio-Services Foundation, Phulwarisharif, Patna

10.1 INTRODUCTION

Phyllanthus emblica L. or Indian gooseberry is a medium-sized tree, belonging to the family Euphorbiaceae, and has found its use as a therapeutic in both traditional and folk medicine (Mirunalini and Krishnaveni 2010). Throughout the world, Ayurveda is considered as one of the oldest traditional systems of medicine (Jaiswal and Williams 2016), and *P. emblica* (*Amlaki* in Sanskrit) is perhaps the most often mentioned plant in Ayurvedic treatise "*Charak Samhita*" (Singh et al. 2011).

The plant is native to India and is also found in China, Sri Lanka, South East Asia, and Uzbekistan (Baliga and Dsouza 2011). All parts of this plant have therapeutic properties owing to the presence of several phenolic compounds such as tannins, phyllembelic acid, phyllemblin, rutin, curcuminoids, emblicol, and other compounds such as gallic acid, quercetin, kaempferol, ascorbic acid, amino acids, flavonoids, and several minerals (Mirunalini and Krishnaveni 2010). Gallic acid has been reported to suppress lung xenograft tumor growth (Ji et al. 2009); quercetin and kaempferol have also shown antineoplastic effects in various cultured cell lines (Nair et al. 2004; Baliga and Dsouza 2011; Devipriya et al. 2008). The globose fruits have found their use in several ailments ranging from the common cold to cancer. They have shown anti-inflammatory, antioxidant, cryoprotective, antidiabetic, hepatoprotective, nephroprotective, and cardioprotective properties (Baliga and Dsouza 2011). These valuable properties have catapulted their role in bearing a possible treatment for cancer, which is what we will explore in this chapter.

10.2 EFFECTIVE CHEMOMODULATOR AND CHEMOPREVENTIVE AGENT

The deleterious effect of chemotherapy on normal cells is a huge price that has to be paid to rid the body of the cancer cells. The side effects usually reflect their mode of action targeting the cell machinery of the normal and tumor cells alike. An agent that can discriminate the healthy cells from the cancerous ones and protect them against the deleterious effects of chemotherapy (chemoprotective) or render the tumor cells more sensitive to anticancer drugs (chemosensitizers) is an exciting proposition in

DOI: 10.1201/9781003477792-10

cancer treatment (Coleman, Bump, and Kramer 1988; Amjad, Chidharla, and Kasi 2022).

Prophylactic treatment with *P. emblica* for a week before the administration of a single dose of thioacetamide (Sultana et al. 2004) or carbon tetrachloride intoxication (Sultana et al. 2005) reverses the drug-induced oxidative stress and early promotional events of primary hepato-carcinogenesis in rats. It inhibited the increase in the levels of serum glutamic oxaloacetic transaminase (SGOT), serum glutamate pyruvate transaminase (SGPT), and γ-glutamyltransferase (GGT); decreased levels of lipid peroxide; inhibited aberrant synthesis of DNA; decreased the activities of glutathione S-transferase (GST), glutathione reductase (GR), glucose-6-phosphate dehydrogenase (G6PD), and ornithine decarboxylase; and concomitantly increased the glutathione content and glutathione peroxidase (GPx) activity in the liver. The acute effects of hepatic toxicity in rat liver could be prevented by pretreatment with *P. emblica* extract.

There are other studies highlighting the prophylactic administration of *P. emblica* extract that has reduced the genotoxic effects of heavy metals and benzopyrene in murine models with bone marrow carcinoma (Dhir et al. 1990; Nandi, Talukder, and Sharma 1997). Continuous administration at 100 mg/kg has also been associated with reduced tumor incidence for skin cancer in mouse by 60% (Sancheti et al. 2005).

What exactly contributes toward the chemopreventive properties of these berries is very difficult to determine. In some species, reactive oxygen species (ROS) may contribute toward a pathological condition and *P. emblica* has been shown to have free radical scavenging activities that might prevent ROS-induced DNA damage and oncogenesis (Hazra et al. 2010; Majeed et al. 2009). The Indian gooseberry has also shown to alter the levels of Phase I and II enzymes (Lu and Xue 2019; Maru et al. 2010; Maru et al. 2016; Shivananjappa and Joshi 2012), thus enabling the bioconversion of carcinogens and purging them from the system. However, there are studies too that report no change in CYP450 (Chen et al. 2011; Tasduq et al. 2005) on *P. emblica* administration, suggesting its radical scavenging properties as a more important contributor. The antitumor (Baliga and Dsouza 2011) and anti-inflammatory (Golechha et al. 2014) properties of *P. emblica* might also be responsible for its chemopreventive nature. It is extremely difficult to dissect out and pin point one specific property or mechanism of action that is responsible for alleviating the course of cancer as this "wonder berry" has a complex mix of compounds (Poltanov et al. 2009; Yang and Liu 2014). It would hence be reasonable to accept that it is the concerted effort of all these mechanisms and molecules that make it so effective.

10.3 EFFECTS ON FREE RADICALS

Oxygen molecules in our body often spontaneously end up with an uneven number of unpaired electrons, forming a highly reactive molecule that damages the lipid cell membrane, proteins, nucleic acid, and other cell constituents (Devasagayam et al. 2004). As a result, oxidative stress plays an important role in the pathogenesis of several diseases like atherosclerosis, alcoholic liver cirrhosis, and cancer. The generation of ROS [ROS superoxide anion radical (O_2^-), hydroxyl radical ($OH^•$), hydrogen peroxide (H_2O_2)], and the reactive nitrogen species [RNS nitric oxide (NO) and peroxynitrite ($ONOO^-$)], in excess, initiates oxidative stress and nitrosative stress (Devasagayam et al. 2004). Since free radicals are naturally produced, the body has antioxidant systems that protect them against oxidative stress by quenching the free radicals (Turrens 2003; Balasaheb Nimse and Pal 2015).

Indian gooseberry has shown the properties for scavenging 2,2-diphenyl-1-picrylhydrazyl (DPPH), superoxide anions, hydroxyl radical, nitric oxide, hydrogen peroxide, peroxynitrite, singlet oxygen, and hypochlorous acid (Naik et al. 2005; Hazra et al. 2010) in different instances of *in vitro* studies. Ethyl acetate fraction of *P. emblica*, containing gallic acid, methyl gallate, corilagin, furosin, and geraniin, showed the highest NO scavenging activity *in vitro* (Kumaran and Karunakaran 2006). In a study on *Bergenia crassifolia* that separated components of the plant extract by TLC, phytochemical components: free gallic and ellagic acids, arbutinemblicanins A and B, hydroquinone, and ascorbic acid were obtained. All these components could scavengeDPPH radicals with emblicanin B being the best scavenger. The scavenging activity decreased in the order: gallic acid > arbutin > ellagic acid > hydroquinone > ascorbic acid (Pozharitskaya et al. 2007). Hence, *P. emblica* extracts may also have the potential to protect cells from oxidative stress-induced diseases such as cancer by preventing ROS-induced DNA damage.

10.4 EFFECTS ON PHASE I AND II ENZYMES AND ON ANTIOXIDANTS

The biotransformation of a drug is its chemical alteration inside a human body. While the liver is the principal site for this, it is not the only organ involved in this metabolism that occurs sequentially in two steps (Lu and Xue 2019). The Phase I reactions involve oxidation, reduction, or hydrolysis of the parent drug, converting it to a more polar molecule by introducing a functional group within the substrate drug molecule. Phase II reactions involve conjugation by coupling the drug or its metabolites to another endogenous molecule, such as acetate, glucuronic acid, glutathione, glycine, and sulfate (Lu and Xue 2019; Wagner, McLaughlin, and Leeder 2020). These reactions further make the metabolized drug more water soluble and thereby enhance its renal excretion.

The cytochrome P450 enzymes (in particular the CYP1, CYP2, CYP3, and CYP4 families) are quantitatively the most important of Phase I, whereas Phase II occurs mainly via methyltransferase, N-acetyltransferase, sulphotransferase, and Uridine 5'-diphospho-glucuronosyltransferase (UDP-glucuronosyltransferase) (Lu and Xue 2019). The concerted activity of Phase I (CYP450) and Phase II enzymes are potentially important for blocking cancer initiation, making metabolizing enzymes a plausible target for cancer chemoprevention (Maru et al. 2010; Maru et al. 2016; Kumar, Pillare, and Maru 2010; Baliga and Dsouza 2011). Studies administering the ethanolic extract of *P. emblica* in Swiss albino mice reveal reduced hepatic levels of the activating enzymes, CYP450 and Cyt b5 in a dose-dependent manner with maximum effect at 500 mg/kg. These enzymes are important in converting the procarcinogen Dimethylbenz(a)anthracene DMBA into a carcinogen (Banu et al. 2004). Administration of *P. emblica* extract (100 μg/mL) on cell lysate of HepG2 cells decreased CYP450 levels while increasing the levels of glutathione (GSH) and enhanced antioxidant defenses of the cells, such as activities of superoxide dismutase (SOD), glutathione peroxidase (GPx), catalase (CAT), GSH reductase (GR), and GSH S-transferase (GST) (Shivananjappa and Joshi 2012). Similarly, in order to test the effect of *P. emblica* extract on the *in vivo* genotoxicity, the extract (250 and 500 mg/kg BW) was orally administered to mice for seven days before giving them a single dose of the mutagens Benzo [α] pyrene (125 mg/kg) or intraperitoneal cyclophosphamide (40 mg/kg). The treatment induced production of GSH and stimulated other antioxidant and detoxification enzymes like GPx, GR, and GST. CYP450 level was also significantly decreased in the animals treated with plant extract (Sharma et al. 2000).

These studies clearly indicate that *P. emblica* extracts not only caused a decrease in the carcinogen-induced transactivation of CYP450 isozymes but also increased the transcription of Phase II enzymes. This modulation of the Phase I and Phase II enzymes, resulting in the decrease of carcinogen-induced DNA adducts, ultimately prevents cancer initiation.

10.5 EFFECT ON INFLAMMATION AND TUMORIGENESIS

Inflammation is associated with many diseases such as cardiovascular and bowel diseases, diabetes, arthritis, and cancer (Libby 2007) and is a part of the process by which the body repairs itself and may be triggered by several signals such as damaged cells, pathogens, and toxic substances (Chen et al. 2017).

Low concentrations of *P. emblica* leaf extract demonstrated anti-inflammatory effects in carrageenan-induced and dextran-induced rat hind paw edema, inhibiting the migration of human polymorphonuclear leucocytes (PMNs) *in vitro*. However, the extract failed to inhibit leukotriene B4 (LTB4) [a leukotriene, produced in leukocytes, involved in inflammation] and platelet-activating factor (PAF) [a potent phospholipid mediator of the many leukocyte functions like inflammation] synthesis within the rat paw model, suggesting that the anti-inflammatory mechanism did not involve lipid mediators (Asmawi et al. 1993). Another study showed reduction in edema and protein content of peritoneal exudates in murine models with two different acute inflammations: carrageenan-induced rat paw edema and acetic acid-induced peritonitis (Dang et al. 2011). Anti-inflammatory effects of *P. emblica* on the tumorigenesis of precancerous lung lesions showed significantly reduced number of nodes on the lung surface. Lung protection was measured by attenuated levels of B(a)P-induced pro-inflammatory cytokines MIP-2, TNF-α, IL-6, and IL-1β in lung tissue, significantly decreased the expressions of COX-2 and HIF-α proteins. Not only this, but the treatment also upregulated the expression of miR-101 while downregulating IL-1β and Lin28B levels (Wang et al. 2017). IL-1β is a pro-inflammatory cytokine (Wang et al. 2014) and Lin28B is a RNA-binding protein, an important pluripotency factor that reprograms translation and promotes cancer establishment (Samsonova et al. 2021). Both these are elevated in the serum of patients with non-small cell lung cancer (NSCLC). microRNA-101 (miR-101), on the other hand, has an established role in tumor suppression. Lin28B is a target of miR-101 and IL-1β represses mir-101, thus resulting in upregulation of Lin28B expression at the posttranscriptional level. Downregulation of mir-101 by IL-1β is crucial for its tumor-promoting activity (Wang et al. 2014). Hence, *P. emblica* extract upregulates mir-101 expression, impeding cancer progression. While most studies look at the phytochemical ingredient, a study evaluated the impacts of oxidative stress and pro-inflammatory markers of biosynthesized silver nanoparticles of *P. emblica* leaves against diethylnitrosamine-induced hepatocellular carcinoma in Wistar rats. A 16-week administration of the biosynthesized nanoparticles significantly enhanced the process of recovery from hepatic cancer. The nanoparticles decreased serum biochemical parameters like alanine transaminase (ALT), aspartate transaminase (AST), alkaline phosphatase (ALP), and alpha fetoprotein (AFP) and altered the level of pro-inflammatory cytokines such as Tumor necrosis factor-alpha (TNF-α), Nuclear factor kappa B (NF-κB), Interleukin-6 (IL-6), and Interleukin-1 (IL-1β) (Singh et al. 2019).

Cancer cells develop the ability to become motile and navigate through the extracellular matrix to infiltrate neighboring tissues, metastasizing on secondary sites. Invasive solid tumors often display collective invasion, in which groups of cells invade the peritumoral stroma while maintaining cell–cell contacts (Friedl et al. 2012). *In vivo* effects of *P. emblica* extracts on mouse xenograft tumor models showed an increase in the expression of E-cadherin by attenuating the migration and invasiveness of carboplatinum- and taxol-resistant high-grade serous ovarian cancer (HGSOC) cell lines (TOV3041G, OV866(2), OV4453, and OV4485). While SNAIL1 and E-cadherin are associated with epithelial–mesenchymal transition (EMT), HIF-1α and IGF1R are associated with angiogenesis. The treatment also significantly reduced the expression of HIF-1α, IGF1R, and SNAIL1 together with reduction in mouse xenograft tumors (>60%, derived from OV4855 cells). The extract therefore exhibited abilities to sensitize platinum- and taxol-resistant heterogenous HGSOC cells with mutations in p53 and BRCA1/2 genes and attenuated their malignant characteristics, suggesting their potential as adjunct therapeutic agents for treating resistant, mutant, and heterogenous ovarian cancer (De et al. 2020).

10.6 EFFECTS ON CELL-CYCLE, CELL PROLIFERATION, AND CELL DEATH

Abnormal proliferation of cells unable to align their growth as predetermined by the cell-cycle mechanism often results in the formation of a tumor. Disruption of the normal progression of the cell cycle and division are significant events in the development of cancer (Kastan and Bartek 2004; Hanahan and Weinberg 2011). This is a tightly regulated process that involves very specific checkpoints to assess extracellular growth signals, cell size along with DNA integrity following precision (Kastan and Bartek 2004). Cyclin and cyclin-dependent kinases (CDKs) play an important role in the progression of the cell cycle, whereas the CDK inhibitors on the other hand prevent the proper functioning of the CDKs, inhibiting the progression of cell cycle. Hence, "activation of positive regulators and functional suppression of negative regulators" determine the genesis of cancer (Hartwell and Weinert 1989; Baliga and Dsouza 2011).

Aqueous extract of *P. emblica* proved cytotoxic with an IC50 (the concentration of an inhibitor at which 50% viability will be achieved is IC50) of 16.5 µg/mL. It also inhibited cell cycle-regulating enzyme cdc25 in a dose-dependent manner with an IC50 of 5 µg/mL, while the concentration needed for 50% inhibition of cdc2 kinase was found to be >100 µg/mL (Jose, Kuttan, and Kuttan 2001). Gallic acid decreases CDKs and cyclins, thereby arresting cell cycle. It phosphorylates Cip1/p21 and cell division cycle 2 (Cdc2), Cdc25A, and Cdc25C in DU145 cells (Sun, Hail, and Lotan 2004). Gallic acid is also known to induce G2/M phase cell cycle arrest via Chk2-mediated phosphorylation of Cdc25C in a bladder transitional carcinoma cell line (TSGH-8301cells) (Ou et al. 2010). *P. emblica* fruits are a rich source of gallic acid, ellagic acid, chebulinic acid, chebulagic acid, emblicanin A, emblicanin B, punigluconin, pedunculagin, citric acid, ellagitannin, trigalloyl glucose, pectin, 1-O-galloyl-β-D-glucose, 3,6-di-O-galloyl-D-glucose, chebulagic acid, corilagin, 1,6-di-O-galloyl-β-D-glucose, 3-ethylgallic acid (3-ethoxy 4,5-dihydroxy benzoic acid), and isostrictiniin (Zhang et al. 2003).

Ovarian cancer is one of the leading causes of female mortality. The extracts of *P. emblica* significantly abated the migration and invasive properties of drug-resistant cell lines of HGSOC phenotypes and significantly reduced the expression of HIF-1α, IGF1R, and SNAIL1 and increased the expression of tumor suppressor protein E-cadherin (De et al. 2020). *P. emblica* extract (50–100 µg/mL) also inhibited the proliferation of cell lines belonging to several other cancers such as that of the human colorectal (SW620), lung (A549), breast (MDA-MB-231), cervical (HeLa), ovarian (SK-OV3), and liver (HepG2) (Ngamkitidechakul et al. 2010).

While cell cycle is responsible for cell duplication, an organism has the capacity to eliminate its unwanted or damaged cells through an evolutionarily conserved, highly ordered mechanism without initiating inflammatory response by apoptosis (Pucci, Kasten, and Giordano 2000). Studies have proved that the process of neoplastic transformation involves successfully disabling the machinery responsible for cell-cycle arrest and apoptosis (Lowe, Cepero, and Evan 2004), bringing down the number of cell deaths in an organism (Sun, Hail, and Lotan 2004). Reinstating apoptosis may effectively eliminate the mutated and severely damaged cells (Collado and Serrano 2010; Kumar et al. 2021) and prove to be a clinically challenging but a potent defense against cancer. *P. emblica* extract can induce apoptosis selectively in certain cell types owing to the presence of certain phytochemicals such as the hydrolysable tannins (Sakagami et al. 2000). Quercetin (Son et al. 2004), gallic acid (Isuzugawa, Ogihara, and Inoue 2001), and pyrogallol (Yang et al. 2009) have also shown cytotoxic and apoptogenic effects on neoplastic and transformed cells but not in normal cells. *P. emblica* extract (50–100 µg/mL) induced apoptosis in HeLa cells by death receptor-mediated apoptosis mechanism by triggering DNA fragmentation, caspase-3, -7, and -8 activity, and expression of Fas protein (Ngamkitidechakul et al. 2010). *P. emblica* extract has also shown to induce apoptosis in Dalton's lymphoma ascites (DLA) and CeHa cell lines at 200 µg/mL. The cell lines also showed that the treatment could also scavenge superoxide and hydroxyl radicals and inhibit lipid peroxidation *in vitro* (Rajeshkumar, Pillai, and Kuttan 2003).

10.7 EFFECTS ON TRANSCRIPTION FACTORS AND ONCOGENES

Transcription factors (TFs) are commonly deregulated in order to establish cancer, so are many more signaling proteins. Interestingly, a specific set of TFs have been seen to be overactive in most human cancer cells, making them targets for the development of anticancer drugs (Darnell 2002). *P. emblica* extract exhibited a dose- and time-dependent inhibition of DNA-binding activity of constitutively active activator protein-1 in both HPV16-positive (SiHa) and HPV18-positive (HeLa) cervical cancer cells. This suggests that *P. emblica* targets the transcription of viral oncogenes responsible for the development and progression of cervical cancer (Mahata et al. 2013), making it a potential plant for the development of anticancer drug.

10.8 CONCLUSION

Although our knowledge about the anticancer properties of *P. emblica* are largely based on the studies on animal models and cell lines and lack *in vivo* human evidence, it still holds its ground as a potential candidate for developing anticancer drug. The biggest advantage is that the fruits of this plant have been consumed by humans for centuries marking its safety. Extensive studies on its various qualities have been compiled here, highlighting its chemopreventative, chemomodulator, and antitumorogenic properties. More research needs to be done to recognize the full potential of this plant in offering a non-toxic drug for the disease that is still a leading cause of death worldwide.

REFERENCES

Amjad, Muhammad T., Anusha Chidharla, and Anup Kasi. 2022. "Cancer Chemotherapy." In *StatPearls*. Treasure Island, FL: StatPearls Publishing. https://www.ncbi.nlm.nih.gov/books/NBK564367/.

Asmawi, Mohd Zaini, Hannu Kankaanranta, Eeva Moilanen, and Heikki Vapaatalo. 1993. "Anti-Inflammatory Activities of Emblica Officinalis Gaertn Leaf Extracts." *Journal of Pharmacy and Pharmacology* 45 (6): 581–84. https://doi.org/10.1111/j.2042-7158.1993.tb05605.x.

Balasaheb Nimse, Satish, and Dilipkumar Pal. 2015. "Free Radicals, Natural Antioxidants, and Their Reaction Mechanisms." *RSC Advances* 5 (35): 27986–6. https://doi.org/10.1039/C4RA13315C.

Baliga, Manjeshwar Shrinath, and Jason Jerome Dsouza. 2011. "Amla (Emblica officinalis Gaertn), a Wonder Berry in the Treatment and Prevention of Cancer." *European Journal of Cancer Prevention: The Official Journal of the European Cancer Prevention Organisation (ECP)* 20 (3): 225–39. https://doi.org/10.1097/CEJ.0b013e32834473f4.

Banu S Mumtaz, Selvendiran K, Singh J Pince Vijeya, Sakthisekaran D. 2004. Protective effect of Emblica officinalis ethanolic extract against 7,12-dimethylbenz(a) anthracene (DMBA) induced genotoxicity in Swiss albino mice. *Human & Experimental Toxicology* 23 (11): 527–31. doi: 10.1191/0960327104ht484oa.

Chen, Kuo-Hsin, Bor-Ru Lin, Chiang-Ting Chien, and Chien-Hsin Ho. 2011. "Emblica officinalis Gaertn. Attentuates N-Nitrosodiethylamine-Induced Apoptosis, Autophagy, and Inflammation in Rat Livers." *Journal of Medicinal Food* 14 (7–8): 746–55. https://doi.org/10.1089/jmf.2010.1459.

Chen, Linlin, Huidan Deng, Hengmin Cui, Jing Fang, Zhicai Zuo, Junliang Deng, Yinglun Li, Xun Wang, and Ling Zhao. 2017. "Inflammatory Responses and Inflammation-Associated Diseases in Organs." *Oncotarget* 9 (6): 7204–18. https://doi.org/10.18632/oncotarget.23208.

Coleman, C. Norman, Edward A. Bump, and Robert A. Kramer. 1988. "Chemical Modifiers of Cancer Treatment." *Journal of Clinical Oncology* 6 (4): 709–33. https://doi.org/10.1200/JCO.1988.6.4.709.

Collado, Manuel, and Manuel Serrano. 2010. "Senescence in Tumours: Evidence from Mice and Humans." *Nature Reviews Cancer* 10 (1): 51–57. https://doi.org/10.1038/nrc2772.

Dang, Geetanjali K., Reshma R. Parekar, Sandhya K. Kamat, A. M. Scindia, and Nirmala N. Rege. 2011. "Antiinflammatory Activity of Phyllanthus emblica, Plumbago zeylanica and Cyperus rotundus in Acute Models of Inflammation." *Phytotherapy Research* 25 (6): 904–8. https://doi.org/10.1002/ptr.3345.

Darnell, James E. 2002. "Transcription Factors as Targets for Cancer Therapy." *Nature Reviews Cancer* 2 (10): 740–49. https://doi.org/10.1038/nrc906.

De, Alok, Archana De, Ramratan Sharma, William Suo, and Mukut Sharma. 2020. "Sensitization of Carboplatinum- and Taxol-Resistant High-Grade Serous Ovarian Cancer Cells Carrying P53, BRCA1/2 Mutations by Emblica officinalis (Amla) via Multiple Targets." *Journal of Cancer* 11 (7): 1927–39. https://doi.org/10.7150/jca.36919.

Devasagayam T. P. A., J. C. Tilak, K. K. Boloor, Ketaki S. Sane, Saroj S. Ghaskadbi, and Rodroclo. Lele. 2004. "Free Radicals and Antioxidants in Human Health: Current Status and Future Prospects." *The Journal of the Association of Physicians of India* 52 (October): 794–804.

Devipriya, Nagarajan, Adluri Ram Sudheer, Marimuthu Srinivasan, and Venugopal P. Menon. 2008. "Quercetin Ameliorates Gamma Radiation-Induced DNA Damage and Biochemical Changes in Human Peripheral Blood Lymphocytes." *Mutation Research* 654 (1): 1–7. https://doi.org/10.1016/j.mrgentox.2008.03.003.

Dhir, Haimanti, Roy Ajoy Kumar, Sharma Archana., and Talukder Geeta. 1990. "Modification of Clastogenicity of Lead and Aluminium in Mouse Bone Marrow Cells by Dietary Ingestion of Phyllanthus emblica Fruit Extract." *Mutation Research* 241 (3): 305–12. https://doi.org/10.1016/0165-1218(90)90029-2.

Friedl, Peter, Joseph Locker, Erik Sahai, and Jeffrey E. Segall. 2012. "Classifying Collective Cancer Cell Invasion." *Nature Cell Biology* 14 (8): 777–83. https://doi.org/10.1038/ncb2548.

Golechha, Mahaveer, Vikas Sarangal, Shreesh Ojha, Jagriti Bhatia, and Dharmveer S. Arya. 2014. "Anti-Inflammatory Effect of Emblica Officinalis in Rodent Models of Acute and Chronic Inflammation: Involvement of Possible Mechanisms." *International Journal of Inflammation* 2014: 178408. https://doi.org/10.1155/2014/178408.

Hanahan, Douglas, and Robert A. Weinberg. 2011. "Hallmarks of Cancer: The Next Generation." *Cell* 144 (5): 646–74. https://doi.org/10.1016/j.cell.2011.02.013.

Hartwell, Leland. H., and Ted. A. Weinert. 1989. "Checkpoints: Controls That Ensure the Order of Cell Cycle Events." *Science (New York, N.Y.)* 246 (4930): 629–34. https://doi.org/10.1126/science.2683079.

Hazra, Bibhabasu, Rhitajit Sarkar, Santanu Biswas, and Nripendranath Mandal. 2010. "Comparative Study of the Antioxidant and Reactive Oxygen Species Scavenging Properties in the Extracts of the Fruits of Terminalia chebula, Terminalia belerica and Emblica officinalis." *BMC Complementary and Alternative Medicine* 10 (May): 20. https://doi.org/10.1186/1472-6882-10-20.

Isuzugawa, Kazuto, Yukio Ogihara, and Makoto Inoue. 2001. "Different Generation of Inhibitors against Gallic Acid-Induced Apoptosis Produces Different Sensitivity to Gallic Acid." *Biological & Pharmaceutical Bulletin* 24 (3): 249–53. https://doi.org/10.1248/bpb.24.249.

Jaiswal, Yogini S., and Leonard L. Williams. 2016. "A Glimpse of Ayurveda – The Forgotten History and Principles of Indian Traditional Medicine." *Journal of Traditional and Complementary Medicine* 7 (1): 50–53. https://doi.org/10.1016/j.jtcme.2016.02.002.

Ji, Bin-Chuan, Wu-Huei Hsu, Jai-Sing Yang, Te-Chun Hsia, Chi-Cheng Lu, Jo-Hua Chiang, Jiun-Long Yang, et al. 2009. "Gallic Acid Induces Apoptosis via Caspase-3 and Mitochondrion-Dependent Pathways *in vitro* and Suppresses Lung Xenograft Tumor Growth *in vivo*." *Journal of Agricultural and Food Chemistry* 57 (16): 7596–7604. https://doi.org/10.1021/jf901308p.

Jose, Jeena K., Girija Kuttan, and Ramadasan Kuttan. 2001. "Antitumour Activity of Emblica Officinalis." *Journal of Ethnopharmacology* 75 (2): 65–69. https://doi.org/10.1016/S0378-8741(00)00378-0.

Kastan, Michael B., and Jiri Bartek. 2004. "Cell-Cycle Checkpoints and Cancer." *Nature* 432 (7015): 316–23. https://doi.org/10.1038/nature03097.

Kumar, Gaurav, Shubhangi P. Pillare, and Girish B. Maru. 2010. "Black Tea Polyphenols-Mediated *in vivo* Cellular Responses during Carcinogenesis." *Mini-Reviews in Medicinal Chemistry* 10 (6): 492–505.

Kumar, Gaurav, Venkateshwar Madka, Gopal Pathuri, Vishal Ganta, and Chinthalapally V. Rao. 2021. "Molecular Mechanisms of Cancer Prevention by Gooseberry (Phyllanthus Emblica)." *Nutrition and Cancer* 74 (7): 2291–2302. https://doi.org/10.1080/01635581.2021.2008988.

Kumaran, Alaganandam, and R. Joel Karunakaran. 2006. "Nitric Oxide Radical Scavenging Active Components from Phyllanthus Emblica L." *Plant Foods for Human Nutrition* 61 (1): 1. https://doi.org/10.1007/s11130-006-0001-0.

Libby, Peter. 2007. "Inflammatory Mechanisms: The Molecular Basis of Inflammation and Disease." *Nutrition Reviews* 65 (12 Pt 2): S140–46. https://doi.org/10.1111/j.1753-4887.2007.tb00352.x.

Lowe, Scott W., Enrique Cepero, and Gerard Evan. 2004. "Intrinsic Tumour Suppression." *Nature* 432 (7015): 307–15. https://doi.org/10.1038/nature03098.

Lu, Jian-Da, and Jun Xue. 2019. "Chapter 101 - Poisoning: Kinetics to Therapeutics." In *Critical Care Nephrology* (Third Edition), edited by Claudio Ronco, Rinaldo Bellomo, John A. Kellum, and Zaccaria Ricci, 600–629.e7. Philadelphia: Elsevier. https://doi.org/10.1016/B978-0-323-44942-7.00101-1.

Mahata, Sutapa, Arvind Pandey, Shirish Shukla, Abhishek Tyagi, Syed Akhtar Husain, Bhudev Chandra Das, and Alok Chandra Bharti. 2013. "Anticancer Activity of Phyllanthus emblica Linn. (Indian Gooseberry): Inhibition of Transcription Factor AP-1 and HPV Gene Expression in Cervical Cancer Cells." *Nutrition and Cancer* 65 (Suppl 1): 88–97. https://doi.org/10.1080/01635581.2013.785008.

Majeed, Muhammed, Beena Bhat, Atul N. Jadhav, Jyotish S. Srivastava, and Kalyanam Nagabhushanam. 2009. "Ascorbic Acid and Tannins from Emblica officinalis Gaertn. Fruits--A Revisit." *Journal of Agricultural and Food Chemistry* 57 (1): 220–25. https://doi.org/10.1021/jf802900b.

Maru, Girish B., Rasika R. Hudlikar, Gaurav Kumar, Khushboo Gandhi, and Manoj B. Mahimkar. 2016. "Understanding the Molecular Mechanisms of Cancer Prevention by Dietary Phytochemicals: From Experimental Models to Clinical Trials." *World Journal of Biological Chemistry* 7 (1): 88–99. https://doi.org/10.4331/wjbc.v7.i1.88.

Maru, Girish, Asha Ramchandani, Gaurav Kumar, and Rachana Garg. 2010. "Curcumin-Mediated Cellular Responses in Chemical Carcinogenesis *in vivo* Studies." In *Bioactive Foods and Extracts*, 181–203. https://doi.org/10.1201/b10330-13.

Mirunalini, Sankaran, and Krishnaveni, Mani 2010. "Therapeutic Potential of Phyllanthus emblica (Amla): The Ayurvedic Wonder." *Journal of Basic and Clinical Physiology and Pharmacology* 21 (1): 93–105. https://doi.org/10.1515/JBCPP.2010.21.1.93.

Naik, G. H., K. I. Priyadarsini, R. G. Bhagirathi, B. Mishra, K. P. Mishra, M. M. Banavalikar, and Hari Mohan. 2005. "*In vitro* Antioxidant Studies and Free Radical Reactions of Triphala, an Ayurvedic Formulation and Its Constituents." *Phytotherapy Research* 19 (7): 582–86. https://doi.org/10.1002/ptr.1515.

Nair, Hari Krishnan, Kesava V. K. Rao, Ravikumar Aalinkeel, Supriya Mahajan, Ram Chawda, and Stanley A. Schwartz. 2004. "Inhibition of Prostate Cancer Cell Colony Formation by the Flavonoid Quercetin Correlates with Modulation of Specific Regulatory Genes." *Clinical and Diagnostic Laboratory Immunology* 11 (1): 63–69. https://doi.org/10.1128/cdli.11.1.63-69.2004.

Nandi, Phalguni, Geeta Talukder, and Archana Sharma. 1997. "Dietary Chemoprevention of Clastogenic Effects of 3,4-Benzo(a)Pyrene by Emblica Officinalis Gaertn. Fruit Extract." *British Journal of Cancer* 76 (10): 1279–83. https://doi.org/10.1038/bjc.1997.548.

Ngamkitidechakul, C., Kanjana Jaijoy, Pintusorn Hansakul, Noppamas Soonthornchareonnon, and S. Sireeratawong. 2010. "Antitumour Effects of Phyllanthus emblica L.: Induction of Cancer Cell Apoptosis and Inhibition of *in vivo* Tumour Promotion and *in vitro* Invasion of Human Cancer Cells." *Phytotherapy Research* 24 (9): 1405–13. https://doi.org/10.1002/ptr.3127.

Ou, Ting-Tsz, Chau-Jong Wang, Yung-Shu Lee, Cheng-Hsun Wu, and Huei-Jane Lee. 2010. "Gallic Acid Induces G2/M Phase Cell Cycle Arrest via Regulating 14-3-3β Release from Cdc25C and Chk2 Activation in Human Bladder Transitional Carcinoma Cells." *Molecular Nutrition & Food Research* 54 (12): 1781–90. https://doi.org/10.1002/mnfr.201000096.

Poltanov, Eugeny A., Alexander N. Shikov, H. J. Damien Dorman, Olga N. Pozharitskaya, Valery G. Makarov, Vladimir P. Tikhonov, and Raimo Hiltunen. 2009. "Chemical and Antioxidant Evaluation of Indian Gooseberry (Emblica officinalis Gaertn., Syn. Phyllanthus emblica L.) Supplements." *Phytotherapy Research: PTR* 23 (9): 1309–15. https://doi.org/10.1002/ptr.2775.

Pozharitskaya, Olga N., Svetlana A. Ivanova, Alexander N. Shikov, Valery G. Makarov, and Bertalan Galambosi. 2007. "Separation and Evaluation of Free Radical-Scavenging Activity of Phenol Components of Green, Brown, and Black Leaves of Bergenia Crassifolia by Using HPTLC-DPPH Method." *Journal of Separation Science* 30 (15): 2447–51. https://doi.org/10.1002/jssc.200700178.

Pucci, Bruna, Margaret Kasten, and Antonio Giordano. 2000. "Cell Cycle and Apoptosis." *Neoplasia (New York, N.Y.)* 2 (4): 291–99.

Rajeshkumar, N. V., M. Radahakrishna Pillai, and R. Kuttan. 2003. "Induction of Apoptosis in Mouse and Human Carcinoma Cell Lines by Emblica officinalis Polyphenols and Its Effect on Chemical Carcinogenesis." *Journal of Experimental & Clinical Cancer Research: CR* 22 (2): 201–12.

Sakagami, H., Y. Jiang, K. Kusama, T. Atsumi, T. Ueha, M. Toguchi, I. Iwakura, et al. 2000. "Cytotoxic Activity of Hydrolyzable Tannins against Human Oral Tumor Cell Lines – A Possible Mechanism." *Phytomedicine: International Journal of Phytotherapy and Phytopharmacology* 7 (1): 39–47. https://doi.org/10.1016/S0944-7113(00)80020-3.

Samsonova, Anastasiia, Krystel El Hage, Bénédicte Desforges, Vandana Joshi, Marie-Jeanne Clément, Guillaume Lambert, Hélène Henrie, et al. 2021. "Lin28, a Major Translation Reprogramming Factor, Gains Access to YB-1-Packaged MRNA through Its Cold-Shock Domain." *Communications Biology* 4 (1): 1–16. https://doi.org/10.1038/s42003-021-01862-3.

Sancheti, Garima, Archana Jindal, Rimpu Kumari, and Pawan Kumar Goyal. 2005. "Chemopreventive Action of Emblica officinalis on Skin Carcinogenesis in Mice." *Asian Pacific Journal of Cancer Prevention: APJCP* 6 (2): 197–201.

Sharma, Neelam, Prashant Trikha, Mohammad Athar, and Sheikh Raisuddin. 2000. "Inhibitory Effect of Emblica officinals on the *in vivo* Clastogenicity of Benzo Alpyrene and Acyclophosphamide in Mice." *Human & Experimental Toxicology* 19 (6): 377–84. https://doi.org/10.1191/096032700678815945.

Shivananjappa, Mahesh Mysore, and Manoj Kumar Joshi. 2012. "Influence of Emblica Officinalis Aqueous Extract on Growth and Antioxidant Defense System of Human Hepatoma Cell Line (HepG2)." *Pharmaceutical Biology* 50 (4): 497–505. https://doi.org/10.3109/13880209.2011.618501.

Singh, Deepika, Ekta Yadav, Neha Falls, Vikas Kumar, Manvendra Singh, and Amita Verma. 2019. "Phytofabricated Silver Nanoparticles of Phyllanthus emblica Attenuated Diethylnitrosamine-Induced Hepatic Cancer via Knock-down Oxidative Stress and Inflammation." *Inflammopharmacology* 27 (5): 1037–54. https://doi.org/10.1007/s10787-018-0525-6.

Singh, Ekta, Sheel Sharma, Ashutosh Pareek, Sachdev Yadav, Swapnil Sharma, and Jaya Dwivedi. 2011. "Phytochemistry, Traditional Uses and Cancer Chemopreventive Activity of Amla (Phyllanthus emblica): The Sustainer." *Journal of Applied Pharmaceutical Science*, 8.

Son, Young-Ok, Kyung-Yeol Lee, Sung-Ho Kook, Jeong-Chae Lee, Jong-Ghee Kim, Young-Mi Jeon, and Yong-Suk Jang. 2004. "Selective Effects of Quercetin on the Cell Growth and Antioxidant Defense System in Normal versus Transformed Mouse Hepatic Cell Lines." *European Journal of Pharmacology* 502 (3): 195–204. https://doi.org/10.1016/j.ejphar.2004.09.012.

Sultana, Sarwat, Salahuddin Ahmad, Naghma Khan, and Tamanna Jahangir. 2005. "Effect of Emblica officinalis (Gaertn) on CCl4 Induced Hepatic Toxicity and DNA Synthesis in Wistar Rats." *Indian Journal of Experimental Biology* 43 (5): 430–36.

Sultana, Sarwat, Salahuddin Ahmed, Sonia Sharma, and Tamanna Jahangir. 2004. "Emblica Officinalis Reverses Thioacetamide-Induced Oxidative Stress and Early Promotional Events of Primary Hepatocarcinogenesis." *The Journal of Pharmacy and Pharmacology* 56 (12): 1573–79. https://doi.org/10.1211/0022357044931.

Sun, Shi-Yong, Numsen Hail, and Reuben Lotan. 2004. "Apoptosis as a Novel Target for Cancer Chemoprevention." *Journal of the National Cancer Institute* 96 (9): 662–72. https://doi.org/10.1093/jnci/djh123.

Tasduq, Sheikh Abdullah, Dilip Manikrao Mondhe, Devinder Kumar Gupta, Meena Baleshwar, and Rakesh Kamal Johri. 2005. "Reversal of Fibrogenic Events in Liver by Emblica officinalis (Fruit), an Indian Natural Drug." *Biological & Pharmaceutical Bulletin* 28 (7): 1304–6. https://doi.org/10.1248/bpb.28.1304.

Turrens, Julio F. 2003. "Mitochondrial Formation of Reactive Oxygen Species." *The Journal of Physiology* 552 (Pt 2): 335–44. https://doi.org/10.1113/jphysiol.2003.049478.

Wagner, Jonathan B., Matthew J. McLaughlin, and Steven Leeder. 2020. "Pediatric Pharmacogenetics, Pharmacogenomics, and Pharmacoproteomics." In *Nelson Textbook of Pediatrics*, 21st ed. Vol. 1. Elsevier. https://www.clinicalkey.com/#!/content/book/3-s2.0-B9780323529501000729.

Wang, Cheng-Cheng, Jia-Rui Yuan, Chun-Fei Wang, Nan Yang, Juan Chen, Dan Liu, Jie Song, Liang Feng, Xiao-Bin Tan, and Xiao-Bin Jia. 2017. "Anti-Inflammatory Effects of Phyllanthus emblica L on Benzopyrene-Induced Precancerous Lung Lesion by Regulating the IL-1β/MiR-101/Lin28B Signaling Pathway." *Integrative Cancer Therapies*. Philadelphia, PA, USA. 16 (4): 505–15. https://doi.org/10.1177/1534735416659358.

Wang, Lin, Ling-Fei Zhang, Jing Wu, Shu-Jun Xu, Yang-Yang Xu, Dangsheng Li, Jia-Tao Lou, and Mo-Fang Liu. 2014. "IL-1β-Mediated Repression of MicroRNA-101 Is Crucial for Inflammation-Promoted Lung Tumorigenesis." *Cancer Research* 74 (17): 4720–30. https://doi.org/10.1158/0008-5472.CAN-14-0960.

Yang, Baoru, and Pengzhan Liu. 2014. "Composition and Biological Activities of Hydrolyzable Tannins of Fruits of Phyllanthus emblica." *Journal of Agricultural and Food Chemistry* 62 (3): 529–41. https://doi.org/10.1021/jf404703k.

Yang, Chih-Jen, Chuan-Sheng Wang, Jen-Yu Hung, Hurng-Wern Huang, Yi-Chen Chia, Pei-Hui Wang, Ching-Feng Weng, and Ming-Shyan Huang. 2009. "Pyrogallol Induces G2-M Arrest in Human Lung Cancer Cells and Inhibits Tumor Growth in an Animal Model." *Lung Cancer* 66 (2): 162–68. https://doi.org/10.1016/j.lungcan.2009.01.016.

Zhang, Lan-Zhen, Wen-Hua Zhao, Ya-Jian Guo, Guang-Zhong Tu, Shu Lin, and Lin-Guang Xin. 2003. "Studies on Chemical Constituents in Fruits of Tibetan Medicine Phyllanthus emblica." *Zhongguo Zhong Yao Za Zhi = Zhongguo Zhongyao Zazhi = China Journal of Chinese Materia Medica* 28 (10): 940–43.

11 Revival of Neglected Crops and Underutilized Crop Cultivation in the Himalayas

Towards Economic Upliftment of Himalayas

Waseem Ali Dar
Krishi Vigyan Kendra/Extension Training Centre, Pulwama, Sher-e-Kashmir University of Agricultural Sciences and Technology of Kashmir, J&K, India

Sajad Majeed Zargar
Proteomics Lab., Division of Plant Biotechnology, Sher-e-Kashmir University of Agricultural Sciences and Technology of Kashmir, J&K, India

J. A. Mugloo
Krishi Vigyan Kendra/Extension Training Centre, Pulwama, Sher-e-Kashmir University of Agricultural Sciences and Technology of Kashmir, J&K, India

11.1 INTRODUCTION

Himalayan ecosystems are one of the youngest and richest ecosystems on earth, covering an area of about 12.84% of the total geographical area of India with varying altitude, topography and climatic conditions (Mani, 1978; Negi, 2009). The northwestern (NW) Himalayas cover about one-fourth of forest land of the world from the middle to upper elevations (Frelich, 2002). Altitudinal variations and climatic and geological conditions in the NW Himalayas contribute immensely to the biological diversity of the mountains (Guangwei, 2002). Soil conditions and rainfall and altitudinal variation of Himalayas generate a variety of distinct plant eco regions, which are well known for its flora and fauna (Mayers *et al.*, 2002) and represent as the natural treasure of biodiversity (Chopra, 2009; Sharma, 2012). The NW Himalayas

DOI: 10.1201/9781003477792-11

approximately comprised 39% grassland, 20% forests, 15% shrub land and 5% agriculture land, and the remaining 21% is made up of barren land, built-up areas, snow cover and water bodies (Pie, 1995; Chettri *et al.*, 2008).

A century of globalization and intensive farming systems in enhancing crop productivity have resulted in the shrinking of diversity in the number of crops and varieties as a diverse group. In the present scenario, our food security depends upon roughly 100,000 to less than 30 crops. Narrowing down the species base at the food security level had restricted livelihood opportunities, which is likely felt by the people living in rural and mountainous areas. The neglected and underutilized crops of today were the major crops of the past. Dietary habits and cropping pattern historical records of tribal and hilly areas of Himalayas indicate that underutilized crops have a distinct position and well-defined role in food and nutritional security at the tribal household level. Indigenous people living in tribal/rural areas possess food resources that are not usually understood by the agriculture and health sector. Underutilized and neglected leafy vegetables being endowed with prophylactic and curative properties are rightly touted as the only source of folic acid, which happens to be critical during pregnancy in humans. The presence of bioactive compounds in the form of phenols, vitamins and omega-3-fatty acid is available in ample amounts in these underutilized vegetables, which makes them good sources of antioxidants. Propagation and significance of these underutilized crops in the tribal/rural community are maintained by their inherited knowledge and conserved by peasants.

Food and Agriculture Organization of the United Nations (FOA) is concerned that by 2050, the global population is expected to reach 9 billion, and the world may not be able to produce enough food to meet the demand. In view of this, FOA and Khon Kaan University organized a symposium in Thailand in 2014 with the objective of raising awareness on the role and value of underutilized indigenous food resources in dietary diversity and household food security and strategic options for the promotion of underutilized indigenous food resources, particularly in the Asia-Pacific region. An effective and sustainable way of development in agriculture and nutritional security depends on broadening the range of neglected and underutilized crops. Today, a vast untapped potential in technology in the field of agriculture is available on hand in underutilized crops. Product development for increasing the tribal/rural income of farmers is a notable feature related to research and extension of underutilized crop species (Bhagmal, 1993; Padulosi *et al.*, 2009). Biodiversity International, International Centre for Underutilized Crops (ICUC) and National Academy of Sciences (NSA, USA) identified 200 underutilized crop species (29 millets, 10 oilseed crops, 27 pulses, 25 root and tuber crops, 52 fruits and nuts, 24 minor fruits, 34 vegetables and 5 fibre and pulp-yielding plants) for different eco-geographical regions of the world (Eyzaguirre *et al.*, 1999).

11.2 STATUS OF NEGLECTED AND UNDERUTILIZED CROPS IN HIMALAYAS

The Indian Council of Agriculture Research started work on the identification and exploration of underutilized crops under the umbrella of IARI, New Delhi, in 1960s

TABLE 11.1
Important Neglected and Underutilized Crop Species of the Himalayan Region

Crop Group	Crop Name
Pseudo-cereals	Buckwheat (*Fagopyrum* spp.), Amaranthus spp. and Chenopodium spp.
Small millets	Barnyard millet (*Echinochola* spp.), finger millet (*Eleusine coracana*), proso millet (*Panicum miliaceum*), pearl millet (*Pennisetum americanum*), little millet (*Panicum sumatrense*) and foxtail millet (*Setaria italica*)
Pulses	Sword bean (*Canavalia* spp.), horse gram (*Lablab purpureus*), horse gram (*Macrotyloma uniflorum*) and faba bean (*Vigna faba*)
Roots and tubers	Yams (*Dioscorea* spp.)
Vegetables	Brassicaceae (*Brassica oleracea* var. *acephala*), Siberan kale, Alliaceae (*Allium cepa* var. *proliferum*) and Amaranthaceae (*Amaranthus viridis*)
	Others: vegetable fern (*Diplazium esculentum*), common purslane/Nunar (*Portulaca oleraceae*), common dandelion/Hannd (*Taraxacum officinale*), common mallow/Sonchal (*Malva sylvestris*), broad-leaved dock/Abuj (*Rumex obtusifolius*), Kamal Kakdi/Nudru (*Nelumbo nucifera*) and stinging nettle/Soye (*Urtica dioica*)
Oil plants	Safflower (*Carthamus tinctorius*) and sesame (*Sesamum indicum*)
Fruits and nuts	Hazelnut (*Corlus avellana*), jackfruit (*Artocarpus heterophyllus*), breadfruit (*A. Allitis*) and pillinut (*Canarium ovatum*)

and was further explored in 1980s by forming All India Coordinated Research Project (ACRIP) on underutilized crops by the National Bureau of Plant Genetic Resources (NBPGR), New Delhi. The ACRIP project on underutilized crops played an important role in collection, preservation and exploration of exotic crops across India (Thakur, 2014). Himalayas represent their own share of neglected and underutilized crops, which are unique with respect to food security, nutritional and medical importance (Gupta *et al.*, 2013; Malik *et al.*, 2021). Generally, underutilized crop species exists as wild and very few are cultivated by indigenous people in their kitchen gardens or in agriculture land. This chapter is meant to compile information on neglected and underutilized crop species from the perspective of their nutritional profile, food security and livelihood opportunities. The important neglected and underutilized crop species identified in the Himalayan region are listed in Table 11.1.

11.3 PSEUDO-CEREALS

Pseudo-cereals are a group of food grains consumed as cereals but are not cereals.

11.3.1 AMARANTHS (*AMARANTHUS* SPP.)

Amaranthus, collectively known as amaranth or pigweed, is a cosmopolitan genus of herb that includes 60 species and belongs to the family Amaranthaceae. Several species are considered as weeds, and a few are used for vegetable and fodder production.

This crop is mainly grown in mid and high hills of the Himalayan region as a pure as well as mixed crop in an area of about 40–50 thousand hectares (Sauer, 1967).

Amaranth is a tinny-seeded annual and fast-growing plant having broad leaves with inflorescence and foliage ranging from purple and red to gold. Small seeds of amaranths are of varying colours and rich in minerals and vitamins and are extremely useful from a human nutrition viewpoint. There are two sections of genus amaranths: Amaranthotypus Dumort (monoecious out crossing species with compound inflorescence) and Blitopsis Dumort (species with large extent of self-pollination). The grain species of amaranthus belongs to Amaranthotypus, which are mostly monoecious, but a few are dioecious. The flowers of amaranths are pentamerous and circumscissile utricle. The basic unit of inflorescence is glomerules, each consisting of an initial staminate flower and an indefinite number of female flowers. The cluster of individual flowers develops along the axis in an alternate fashion (Joshi and Rana, 1991). Most of the species have $2n=32$ or 34, while the polyploid species *Amaranthus dubius* has $2n=64$ (WOI, 1985).

Amaranth has very high nutritive value, rich in protein, beta-carotene and folic acid, and helps in increasing the blood haemoglobin level. Amaranths contain lysine protein more than twice that of wheat and thrice that of maize. Amaranth grains contain 14%–19% crude protein, 3.1% fat, 0.49% calcium, 0.60% phosphorus and 0.14% iron; in addition, amaranth grains possess 8% dietary fibre (Saunders and Becker, 1983; Joshi and Rana, 1991; Solanki *et al.*, 2008).

Aamaranth can foster rural life and has the potential to improve nutrition and boost food security and sustainable land care. It is an inexpensive native crop that could be cultivated by indigenous rural/tribal people due to its nutritive value and is tolerant to arid environments. Aamaranth has multipurpose utility as food, fodder, medicine and industrial oil and secures livelihood opportunities (Sanchez-Marroquin, 1983; Ahuja et al., 1991; Phogat and Sharma, 2000).

Amaranth has great potential in agro-based industries for application in high-quality plastics, cosmetics, pharmaceuticals and natural dyes. The grains from amaranths are used in the preservation of meat and apple fruit. Tribal people of the Himalayas in India use amaranths grains for the treatment of measles and snake bites in humans, and for foot and mouth diseases in animals. Stem and leaf extracts of amaranthus are used for the treatment of kidney stones. The topopherol fraction of amaranth oil contains important cholesterol-lowering agents. In some parts of the Indian Himalayas, amaranth seeds are fermented to make beer and tonics (Gupta *et al.*, 2013).

11.3.2 Buckwheat

Buckwheat (*Fagopyrum* spp.), a member of the dicot family polygonaceae, is widely distributed in the Indian Himalayas at varying frequencies (Ye and Guo, 1992; Farooq, 2016). There are two cultivated species of buckwheat, namely *Fagopyrum esculentum* Moench called as common buckwheat and *F. tataricum* Gaertn., known as tartary buckwheat. Buckwheat is a cross-pollinated and self-sterile erect annual herb with a hallow stem, swollen nodes and alternate triangular leaves. In high-altitude areas, buckwheat fits well in the farming system because of short-duration crops and requires less care in crop raising (Joshi and Paroda, 1991). Buckwheat thrives best in sandy soils under harsh conditions and is sensitive to high temperatures and hot dry winds but tolerant to acid soils.

Buckwheat is a multipurpose crop used in a variety of ways. In Nilgiri Hills, buckwheat is used as green manure and its cultivation can improve the physical condition and water-holding capacity of the soil. In the mid-hills of the Indian Himalayas, buckwheat tender shoots are used as leafy vegetables.

Buckwheat is a pseudo-cereal and its grains are highly nutritious in protein (15%), fat (5%) and a good source of iron 155 mg/100 g (Lyman *et al.*, 1956; Skrabanja *et al.*, 2004). It is an excellent quantity of rutin, an antioxidant compound, and is also a good source of lysine (Halosava *et al.*, 2002; Dutta, 2004). Consumption of buckwheat products on a regular basis will reduce the risk of high blood pressure and arteriosclerosis and prevent edema and haemorrhagic diseases (Kim *et al.*, 2004; Prestamo *et al.*, 2003). The flower of common buckwheat produces very good-quality honey.

11.4 MINOR MILLETS

Millets represent a diverse group of small-seeded grasses grown for food, feed or forage and are sometimes referred to as orphan crops or lost crops (Lata, 2015). However, these neglected crops are important by virtue of their contribution to food security and livelihood opportunities (Belton and Taylor, 2004). In India, six minor millet species, viz., foxtail millet, proso millet, finger millet, kodo millet, little millet and barnyard millet, are cultivated, and all of them are known for their unique traits and nutritional value in India. Millets need little water and can be cultivated under non-irrigated conditions or in low rainfall regimes (200–250 mm). Finger millet (ragi) is the most popular minor millet in the Himalayan region among others and is cultivated in an area of around 1.6 million ha, with an annual production of 2.4 million tonnes and a productivity of approximately 1,534 kg/ha (Government of India, 2009). The rest of the other millets are cultivated in an area of 1.1 m ha, with a production of 0.7 million tonnes. Various research and developmental programmes are required to improve and optimize the diversified utilization of minor millets in India.

11.4.1 Finger Millet (*Eleusine coracana*)

The name English finger millet is derived from the finger-like structure of the panicle. The generic name *Eleusine* is derived from the Greek goddess of cereals, '*Eleusine*', and ranks first in importance among millets in India. Finger millet is highly adaptable up to an altitude of 2,300 m and can thrive under a variety of harsh environmental conditions and can be stored for 10 years, which makes it a perfect food grain commodity for famine-prone areas in the Himalayas (National Research Council, 1996; Gull *et al.*, 2014).

Finger millet is an erect tetraploid tufted annual plant ($2n = 4x = 36$; genome constitution of AABB) growing 60–120 cm in height with a fibrous and adventitious root system. Plant tillers profusely with compressed stems, and the leaves are linear with distinct midrib, ligule and fringe of hairs. Finger-like ear heads are produced at the end of tillers, which bear crowded/overlapped spikelets. Each spikelet contains four to five flowers and may take six to eight days to complete flowering.

Finger millet is an important source of diet for pregnant ladies, lactating mothers and children, as well as for the economy of marginal farmers. Its grain contains more fibre, minerals and vitamins, which are normally deficient in the Indian diet. Finger millet grain constitutes nearly 80% carbohydrate and contains relatively less fat and protein content than major cereals (Gupta *et al.*, 2013; Kumar *et al.*, 2014). Its seed coat is rich in phytochemicals like dietary fibre and polyphenols and it also contains a high concentration of calcium and soluble fibre and low in glycemic index of grains (Seethram and Rao, 2007; Davi *et al.*, 2014). Finger millet grains are superior to cereals, especially to rice and wheat, as they have a good number of essential amino acids such as valine (413 mg/g of protein), isoleucine (275 mg/g of protein), threonine (263 mg/g of protein), methionine (194 mg/g of protein) and tryptophan 191 mg/g of protein (Fernandez *et al.*, 2013; Gupta *et al.*, 2013).

Finger millet has the property of low digestibility, thereby furnishing energy throughout the day. It is reported to be diaphoretic, diuretic and vermifuge, and its leaf juice has been given to women in childbirth (Dida and Devos, 2006). Finger millet has been used in the preparation of traditional foods in the Himalayas since time immemorial. It forms a staple diet in many tribal and rural households and fetches good sustainable livelihood security (Tables 11.2 and 11.3).

TABLE 11.2
Nutritional Status of Minor Food Crops per 100 g (Joshi and Paroda, 1991)

Food Grain	Protein (g)	Carbohydrates (g)	Lipid (g)	Crude Fibre (g)	Calcium (g)	P (g)	Fe (g)
Amaranth	16.0	62.0	8.0	2.43	490	600	17.5
Buckwheat	13.0	72.9	7.4	10.5	120	280	15.5
Foxtail millet	12.3	60.9	4.3	4.3	31	290	5.0
Proso millet	12.0	60.0	4.0	4.1	30	288	5.0
Wheat	11.0	69.0	1.7	1.2	41	306	5.3
Rice	6.7	69.0	0.3	0.2	45	160	3.5
Milk	5.8						

TABLE 11.3
Essential Amino Acid Composition (g/100 g Protein) in Minor Crops in Comparison to Cereals and Milk (Bhagmal, 1994)

Crop	Lysine	Cystine	Methionine	Isoleucine	Leucine
Amaranth	5.0	4.0	4.0	3.0	4.7
Buckwheat	6.2	1.6	1.6	3.7	6.2
Foxtail millet	2.2	1.6	2.8	7.6	16.7
Proso millet	3.0	1.0	2.6	8.1	12.2
Wheat	2.8	2.2	1.5	3.3	6.7
Rice	3.8	1.4	2.3	3.8	3.2
Milk	5.8	2.1	3.7	5.0	7.3

11.5 MINOR PULSES

In agriculture, legumes have special significance by providing a sustaining source to other crops through enrichment by symbiotic nitrogen fixation and also add value to human diets by providing a high-protein source. Besides the major pulses, there are other several legumes which are cultivated for multifarious uses in marginal lands in remote areas and have tremendous scope to become an important component of the agricultural system. These underutilized legumes are important because of their intrinsic resilience and have useful traits. Underutilized crops have multiple edible parts: flowers, seeds, pod case, immature pods, tuberous roots and leaves and give a chance for the human nutritional requirement to be satisfied through different diets (Mohanty *et al.*, 2020; Ibirinde *et al.*, 2019). Many underutilized legumes are very rich in nutrients and can promote good health and ensure food security (Ntaganda, 2022). Plant-based proteins are very affordable, especially in this era where animal protein sources pose a health threat and are being discouraged (Cheng, 2018). This led to the promotion and encouragement of plant-based protein to fullfill the nutritional needs of the ever-growing population (Cheng *et al.*, 2019). The most important ones grown in the NW Himalayas are rice bean, faba bean, winged bean and adzuki bean.

11.5.1 Rice Bean (*Vigna umbellata*)

Rice bean, also known as the climbing bean, mountain bean, oriental bean and haricot bean, is a highly branched annual with an erect or semi-erect stem tending to viny. Rice bean has raceme-type inflorescence with bright yellow-coloured flowers and slender or curved pods. It is largely a self-pollinated diploid ($2n=22$) crop, but some natural cross-pollination has also been reported (Sastrapradja and Sutarno, 1977; Singh and Sharma, 1993).

Rice bean is a multipurpose grain legume crop cultivated for food, fodder and green manure resources by farmers in hilly areas and is mainly grown between 700 and 1,300 m above sea level (Gautam *et al.*, 2007). It is widely used as an intercrop, particularly for maize (Lawn, 1995).

Rice beans mainly contain minerals, essential amino acids, high protein content and vitamins. Low fat content and a relatively high proportion of healthy unsaturated fatty acids make it a nutritional health package. Rice bean provides an ecological service because of its high nitrogen content. Besides enriching soil nitrogen, it also increases a good amount of humus in the soil, which helps in improving the soil structure porosity and also contributes to control weeds (Table 11.4).

11.6 UNDERUTILIZED VEGETABLES

Indigenous vegetables are replaced by a few cultivated species of which may be lost due to neglect. In India, vegetables are grown from dry temperate to humid tropics between the altitudes from sea level to snow line. Himalayan and sub-Himalayan regions in India constitute an area where a wide range of plant species belonging to the genus Solanaceae, Cucurbitaceae and Brassicaceae are grown. Besides these,

TABLE 11.4
Nutritive Value (per 100 g) of Underutilized and Major Food Legumes (Bhagmal, 1994)

Crop	Crude Protein (g)	Fat (g)	Carbohydrate (g)	Fibre (g)	Ash (g)
Rice bean	20.9	0.9	60.7	4.8	4.2
Faba bean	26.2	1.3	59.4	6.8	3.0
Adzuki bean	19.9	0.6	64.4	7.8	4.3
Winged bean	32.8	17.0	36.5	4.1	3.6
Chickpea	19.4	5.6	60.9	2.5	3.1
Pigeon pea	21.6	1.4	72.7	8.1	4.2

various cole crops, tuber crops, legumes and a few species of leafy vegetables constitute a long list of underutilized vegetables. These vegetables are generally eaten by the urban/tribal population and are rarely shipped out as commercial commodities. Some of these vegetables are not cultivated generally but are gathered from the wild. These traditional/underutilized leafy vegetables are highly nutritious and are also important in food security during times of drought and vital for income generation.

Withstanding their value as food, these vegetables also serve as a source of medicines and are also important in cultural values. The most important ones grown in the NW Himalayas are spear-leaves yam, vegetable fern, goosefoot, horse purslane, Malabar spinach, Indian lotus, Kale, Sonchal sag, Shallot, Sorrel and Pulsane, which have medical significance and multifarious uses in Tables 11.5 and 11.6 (Tsering and Bushin, 2022).

1. **Vegetable fern** (*Diplazium esculentum*): Diplazium is a mostly used underutilized vegetable found in NW Himalayan states. Its young leaves are cooked and also used as salad or pickle. Decoctions of young leaves are used as a tonic for lactating women, as it contains antioxidant vitamin A and carotene. Its aerial parts are used to treat haemoptysis and cough; rhizomes are used for diarrhoea, dysentery and coughs; leaves are used to treat fever and dermatitis (Tsering and Bushin, 2022).
2. **Kale** (*Brassica oleraceae* var. *acephala*): It is mainly distributed in the Northern Himalayan region, and in Kashmir, it is one of the oldest cultivated vegetables and is popularly grown as '**Hak/Saag**'. It is hardy, cool season and withstands against freezing temperatures between −10°C and −15°C. Its green leaves are cooked and also used as salad. The health benefits of kale are well known as it is an excellent source of antioxidant food and a rich source of vitamin K, ascorbic acid, provitamin A, vitamin B6, carotenoids and omega-3 fatty acids and is called a superfood (Armesto *et al.*, 2015). It serves as a powerful tool for heart protection, blood pressure lowering, anti-inflammatory anti-viral, anti-depressant and anti-cancer effects (Tsering and Bushin, 2022).

TABLE 11.5
Species, Common Name and Botanical Features of Underutilized Vegetables of Himalayan Region

Common Name	Bot. Name	Local Name	Family	Part Used
Spear-leaved yam	*Dioscorea belophylla*	Tarar	Dioscoreaceae	Bulbs
Horse purslane	*Trianthema portulacastrum*	Chaulai	Amaranthaceae	Leaves and seeds
Vegetable fern	*Diplazium esculentum*	Kasrod	Polypodaceae	Leaves
Goosefoot	*Chenopodium album* Linn	Bathua	Chenopodeceae	Tender leaves and stem
Malabar spinach	*Basella* spp.	Poi Saag	Basellaceae	Leaves
Indian lotus	*Nelumbo nucifera*	Nadru	Nelumbonaceae	Lotus stems
Kale	*Brasica oleraceae* var. *acephala*	Haak	Brassicaceae	Leaves and stem
Sonchal Sak	*Malva sylvestris* Linn	Sonchal	Malvaceae	Leaves and Stem
Shallot	*Allium ascalonicum*	Pran	Alliaceae	Bulbs
Sorrel	*Rumex acetosa*	Abej	Ploygonaceae	Leaves
Purslane	*Portulaca olreaceae*	Nuner	Portulacaceae	Leaves

TABLE 11.6
Nutritive Value (per 100 g Edible Portion) of Major Underutilized Vegetables Grown in Northern Himalayan Region (Tsering and Bushin, 2022)

Common Name	Protein	Fat	Carbohydrates	Calcium	Phosphorus	β–Carotene	Iron	Vit. C
Spear-leaved yam	1.53 g	0.1 g	12.8 g	19 mg	38 mg	83 g	0.54 mg	–
Horse purslane	2.6 g	0.5 g	5.46 g	52. 6 mg	–	4.56 mg	10.3 mg	–
Vegetable fern	2.5 g	0.20 g	5 g	22 mg	62 mg	449 mg	10.2 mg	24.4 mg
Goosefoot	3.0 g	0.7 g	2.0 g	200 mg	50 mg	1,740 μ g	1.2 mg	–
Malabar spinach	1.8 g	0.3 g	3.4 g	106 mg	52 mg	–	1.48 mg	–
Indian lotus	1.58 g	0.07 g	17 g	16.02 mg	100 mg	–	1.16 mg	27.4 mg
Kale	3.9 g	0.6 g	6 g	177 mg	49.6 mg	2,040 mg	2.5 mg	21 mg
Sonchal Sak	4.3 g	0.6 g	5.6 g	300 mg	60 mg	2,490 μ g	19.5 mg	79 mg
Shallot	1.8 g	0.1 g	17.6 g	50 mg	70 mg	–	2.3 mg	8 mg
Sorrel	2.6 g	0.7 g	2.6 g	43 g	63 mg	266 μ g	3.2 mg	48 mg
Purslane	2.4 g	0.1 g	2.9 g	65 mg	44 mg	1,320 IU	1.99 mg	26.6 mg

11.7 UNDERUTILIZED CROPS: TRENDS, CHALLENGES AND LIVELIHOOD OPPORTUNITIES IN THE 21ST CENTURY

Plant biodiversity on earth represents the primary source of food and feed, medicines and many other products that make life possible and enjoyable. Only one-third of the plant species used by humans generate diverse cultures around the world that would meet specific needs. The increased reliance on major food crops has been accompanied by the shrinking of the food basket which humankind has been relying on for generations. The 20th century witnessed the systematic collection of staple food crops to rescue genetic resources, but the 21st century has started with the awareness of the need to rescue and improve the use of neglected and underutilized crops that are left aside by research, modern technologies and marketing systems as well as conservation efforts (Pistorius, 1997). Policymakers have included these neglected and underutilized crops in worldwide action plans after having raised interest. In 1999, the Global Forum on Agricultural Research (GFAR) emphasized the role of underutilized species in raising the income of the rural poor (Frison *et al.*, 2000).

Future progress of underutilized crops in the domestication for integrated rural development will probably come with innovative processing technologies. The ideotype approach will be needed to target products for wider range of commercial markets that sophisticated domestication of underutilized crops.

11.8 POLICIES TO BE IMPLEMENTED TO PROMOTE UNDERUTILIZED CROPS

As cultivation and use of underutilized crops would address the sustainable development goals of ending hunger, poverty alleviation, health, climate change and sustainable use of ecosystems, governments need to accord adequate attention, priority and funding for its development and mainstreaming. 'Steering Committees' need to be constituted at the national level, comprising policymakers, researchers, Non-Government Organization (NGOs)/ Non-Profit Organization (NPOs), private industry players (especially for value addition) and farmer's representatives. 'Working Groups' and 'Technical Groups' may be established at the national and regional levels. The Chairs of these working groups should be members of the Steering Committees to develop a national policy on underutilized crops. Policy-based interventions like food subsidy, cultivation subsidy and promotion of government food schemes (e.g. mid-day meals, promotion of multigrain flour) would significantly promote underutilized crops and address the welfare of the poorest of the poor.

REFERENCES

Ahuja AK, Multani KK, Gupta BK, and Batta RK (1991). Studies on the chemical composition and dry matter digestion kinetics of amaranths. *SARAS J Livest Poult Prod* 61(12): 15–17.

Armesto J, Carballo J, and Martinez S (2015). Physicochemical and phytochemical properties of two phenotypes of Galegalale (*Brassica oleracea* L. var *acephala* cv. Galega). *J Food Biochem* 39(4): 439–448.

Belton PS and Taylor JRN (2004). Sorghum and millets: protein sources for Africa. *Trends Food Sci Technol* 15: 94–98.

Bhagmal S (1993). Genetic resources of underexploited plants. In: *Exploration, evaluation and conservation of plant genetic resources*. (RS Rana, B Singh, UC Srivvastava, PN Mathur, M Rai, BM Singh, S Kochar, and IS Bisth (eds)). National Bureau of Plant genetic Resources, New Delhi, India, 158–164 pp.

Bhagmal (1994). *Underutilized Grain Legumes and Pseudocereals- their Potential in Asia.* Regional offices for Asia and the Pacific (RAPA), FAO, Bangkok, Thailand, 162 p.
Cheng A (2018). Shaping a sustainable food future by rediscovering long-forgotten ancient grains. *Plant Sci* 269: 136–142.
Cheng A, Raai MN, Zain NAM, Massawe F, Singh A, and Wan WAAQI (2019). In search of alternative proteins: unlocking the potential of underutilized tropical legumes. *Food Secur* 11(6): 1205–1215.
Chettri N, Shakya B, Thapa R, and Sharma E (2008). Status of a protected area system in the Hidu Kush-Himalayas: an analysis of PA coverage. *Int J Biodivers Sci Manag* 4: 164–178.
Chopra AK (2009). Impact of climate change on biodiversity of India with special reference to Himalayan region: an overview. *Appl Nat Sci Found* 1: 117–122.
Davi PB, Vijayabharathi R, Sathyabama R, Mallesh NG, and Priyadarisini VB (2014). Health benefits of finger millet (*Eleusine coracana* L.) polyphenols and dietary fibre: a review. *J Food Sci Technol* 51: 1021–1040.
Dida MM and Devos KM (2006). Finger millet. In: *Cereals and millets.* (C Kole (ed)). Springer, New York, 333–343 pp.
Dutta M (2004). Buckwheat improvement in India: current status and future prospects. In: *Proceedings of the 9th International Symposium on Buckwheat, Progue*, 302–312 pp.
Eyzaguirre PB, Padulosi S, and Hodgkin T (1999). IPGRI'S strategy for neglected and gunderutilized species and the human dimensions of agro-biodiversity. In: *Priority setting for underutilized and neglected plant species of the Mediterrean region.* (S Padulosi (ed)). International Plant Genetic Resources Institute, Rome, Italy, 1–9 pp.
Farooq S (2016). Cultivation, agronomic practices, and growth performance of buckwheat. In: *Molecular breeding and nutritional aspects of buckwheat.* (Meiliang Z (ed)). Academic Press, Gunila, Wieslander, 299–319 pp.
Fernandez DR, Vanderjagt DJ, Millson M, Huang YS, Chuang LT, and Pastuszyn A (2013). Fatty acid, amino acid and trace mineral composition of Eleusine coracana (Pwana) from seeds from northern Nigeria. *Plant Foods Hum Nutr* 58: 1–10
Frelich LE (2002). *Forest dynamic and disturbance regimes studies from temperate evergreen deciduous forests.* Cambridge University Press, Cambridge, UK.
Frison EH, Omont H, and Padulsoi S (2000). GFAR and International cooperation on commoditychains. In: *Synthesis Paper Presented at the GFAR-2000 Conference*, held in Dresden, Germany on 21–23 May.
Gautam R, Kumar N, Yadavendra, JP, Neog SB, Thakur S, Khanal A, Bhandari B, and Hollington PA (2007). *Food security through ricebean research in India and Nepal (FOSRIN). Report on distribution of ricebean in India and Nepal.* Local Initiatives for Biodiversity, Research and Development, Pokhara, Nepal and CAZS Natural Resources, College of Natural Sciences, Bangor University, Wales, UK.
Government of India (2009). *Area, Production and Productivity of Millets in India.* Directorate of Millets, Ministry of Agriculture, Jaipur, India.
Guangwei C (2002). Biodiversity in the Eastern Himalayas: conservation through dialogue. In: *Summer Reports of the Workshops on the Biodiversity Conservation in the Hindu Kush-Himalayan Ecoregion.* ICIMOD, Kathamandu.
Gull A, Jan R, Nayik GA, Prasad K, and Kumar P (2014). Significance of finger millet in nutrition, health and value added products: a review. *J Environ Sci Comput Sci Eng Technol* 3: 1601–1608.
Gupta A, Sood S, Agarwal PK, and Bhatt JC (2013) Underutilized food crops of Himalayan region: utilization and prospective. In: *Newer approaches to biotechnology.* (KK Behera (ed)). Narendra Publishing House, Delhi, India, 101–120 pp.

Halosava M, Fiedlerova V, Smrcinova H, Orsak M, Lachman L, and Vavreinova S (2002). Buckwheat- the source of antioxidant activity in functional foods. *Food Res Int* 35: 207–211.

Ibirinde DO, Aremu CO, Balogun K, and Oladokum L (2019). Assessment of seed and tuber production potential in varities of *Sphenostylis stenocarpa* (Africa Yam Bean). *Agric Sci* 10: 870–881.

Joshi BD and Paroda RS (1991). *Buckwheat in India.* National Bureau of Plant Genetic Resources, New Delhi, India.

Joshi BD and Rana RS (1991). Grain Amaranth: the future food crop. *NBPGR Shimla Sci Monogr* 3: 152.

Kim SL, Kim SK, and Park CH (2004). Introduction and nutritional evaluation of buckwheat sprouts as a new vegetable. *Food Res Int* 37: 319–327.

Kumar A, Mirza N, Charan T, Sharma N, and Gaur VS (2014). Isolation, characterization and immunolocalization of a seed dominant CaM from finger millet (*Eleusine coracana* L Gartn.) for studying its functional role in differential accumulation of calcium in developing grains. *Appl Biochem Biotechnol* 1172: 2955–2973.

Lata C (2015). Advances in omics for enhancing abiotic stress tolerance in millets. *Proc Indian Nat Natl Sci Acad* 81:397–417.

Lawn RJ (1995). The Asiatic Vigna species. In: *Evolution of crop plants.* (J Smartt and NW Simmonds (eds)). Longman Scientific and Technical, Harlow, UK, 321–326 pp.

Lyman CM, Kulken KA, and Hall F (1956). Essential amino acid content of farm feeds. *J Agric Food Chem* 4: 1008–1013.

Malik G, Singh BD, Malik AJ, and Ariza G (2021). *Underutilized vegetable species of Indian Himalayas.* In: *Neglected and Underutilized Crops-Towards Nutritional Security and Sustainability.* (Zargar SM (ed)). Springer Verlag, Singapore, 113–132 pp.

Mani MS (1978). *Ecology and phytogeography of the high altitude plants of the Northwest Himalaya: introduction to high altitude botany.* Hasstead Press, Ultimo, Australia, 205 p.

Mayers N, Mittermeier RA, Mittermeier CG, Fonseca D, and Kent GAB (2002). Biodiversity hotspots for conservation priorities. *Nature* 403(**24**): 853–858.

Mohanty CS, Singh V, and Chapman MA (2020). Winged Bean: an underutilized tropical legume on the path of improvement to help mitigate food and nutrition security. *Sci Hortic* 260: 108789.

National Research Council (1996). *Lost crops of Africa. Vol. 1.* Grains. National Academy Press, Washington DC, 383 p.

Negi SP (2009). Forest cover in Indian Himalayan states - an overview. *Indian J For* 32: 1–5.

Ntaganda J (2022). Analysis of risk factors that influence stunting among Rwandan children under the age of five. *Afr J Food Agric Nutr Dev* 22(**5**): 1–10.

Padulosi S, Mal B, Ravi SB, Gowda J, Gowda KTK, Shanthakumar G, Yenagi N, and Dutta M (2009). Food security and climate change: role of Plant Genetic Resources of Minor Millets. *Indian J Plant Genet Resour* 22(**1**): 1–6.

Phogat BS and Sharma GD (2000). *Underutilized food crops: their uses, adaptation and production technology.* National Bureau of Plant Genetic Resources, New Delhi, India, 58 p.

Pie S (1995). Banking on biodiversity. In: *Report on the Regional Consultations on Biodiversity Assessment in the Hindu-Kush-Himalaya.* ICIMOD, Kathmandu.

Pistorius R (1997). *Scientists, plants and pollitics-a history of the plant genetic resources movement.* International Plant Genetic Resources Institute, Rome, Itlay, 134 p.

Prestamo G, Pedrazuela A, Penas E, Lasuncion MA, and Arroyo G (2003). Role of buckwheat diet on rats as prebiotic and healthy food. *Nutr Res* 23: 803–814.

Sanchez-Marroquin A (1983). Two forgotten crops of agro-industrial importance: Amaranth and Quinoa. *Arch Latinoam Nutr* 33: 11–32.

Sastrapradja S and Sutarno H (1977). Vigna umbellate (L.) DC, in Indonesia. *Ann Bogor* 5(**3**): 155–167.

Sauer JD (1967). Then grain amaranths and their relatives: a revised taxonomic and geographic survey. *Ann Missouri Bot Gard* 54(**2**): 103–107.

Saunders RM and Becker R (1983). Amaranthus. In: *Advances in cereal science and technology*. (Y. Permerenj (ed)). American Association of Cereal Chemistry, St. Paul, MN, 357 p.

Seethram N and Rao BD (2007). *The Hindu Survey of Indian Agriculture*. 36–38 pp.

Sharma R (2012). Impacts on human health of climate and land use change in the Hindu Kush Himalayan region source. *Mt Res Dev* 32: 480–486.

Singh M and Sharma BK (1993). Natural outcrossing in rice bean. *Indian J Agric Sci* 63: 530–521.

Skrabanja V, Kreft I, Golob T, Modic M, Ikeda S, and Ikeda K (2004). Nutrient content in buckwheat milling fractions. *Cereal Chem* 81(**2**): 172–176.

Solanki SD, Ravindrababu Y, Patel BM, Patel AM, Joshi V, and Tikka SBS (2008). *Status of underutilized crops in Gujarat*. AICRN on Underutilized Crops, Regional Research Station, SD Agricultural University, Sardarkrushinagar, Gujarat.

Thakur M (2014). Underutilized food crops: treasure for the future India. *Food Sci Res J* 5(**2**): 174–183.

Tsering L and Bushin B (2022). Indigenous vegetables resources and their utilization under North Western Himalayan Region. *Int J Emerg Appl Sci Technol* 2(**7**): 374–380.

WOI (1985). *The wealth of India-raw materials*. Vol. 1: Publications and Information Directorate, CSIR, New Delhi, India.

Ye N and Guo G (1992). Classification, origin and evolution of genus Fagopyrum in China. In: *Proceedings of 5th International Symposium on Buckwheat*, 20–26 August 1992, Taiyuan, China, 19–28 pp.

Index

Note: **Bold** page numbers refer to tables; *italics* page numbers refer to figures.